Green Energy and Technology

More information about this series at http://www.springer.com/series/8059

Noorhana Yahya

Green Urea

For Future Sustainability

Springer

Noorhana Yahya
Universiti Teknologi PETRONAS
Seri Iskandar, Perak
Malaysia

ISSN 1865-3529 ISSN 1865-3537 (electronic)
Green Energy and Technology
ISBN 978-981-10-7577-3 ISBN 978-981-10-7578-0 (eBook)
https://doi.org/10.1007/978-981-10-7578-0

Library of Congress Control Number: 2017961099

Printed on acid-free paper

This Springer imprint is published by Springer Nature
The registered company is Springer Nature Singapore Pte Ltd.
The registered company address is: 152 Beach Road, #21-01/04 Gateway East, Singapore 189721, Singapore

Acknowledgements

I would like to extend my sincere gratitude to the Ministry of Higher Education (MOHE) for acknowledging Green Technology by providing a RM 12 million worth grant for the 'Green and Economical Urea' project under the Long Term Research Grant (LRGS) program. The publication of this book is a collaborative effort of all researchers under the LRGS program, especially to Assoc. Prof. Dr. Ku Zilati KuShaari, Prof. Dr Khanif, AP Dr. Shahrina, Assoc. Prof. Dr. Zainovia, Prof. Dr. Azizah and Assoc. Prof. Dr. Zakaria Man and I sincerely thank each one of them for the countless support. I would also like to express my appreciation to my Postdoctoral researcher, Dr. Menaka Ganeson and all my Post Graduate students, Dr. Saima Qureshi, Dr. Bilal Alqasem, Mr. Zia Ur Rehman and Mr. Muhammad Irfan for conducting the research and their support in completing the book. I am eternally grateful to Prof. Mike Payne, University of Cambridge and Prof. Krzysztof Koizol, University of Crainfield for their valuable comments and constructive feedbacks in completing this book. Last but not least, I wish to thank my husband, Sazali, my daughter Aalia and my son, Zikri for giving me the space of time and for the endless support and love.

Noorhana Yahya
Universiti Teknologi PETRONAS
2018

Contents

About the Author

Prof. Dr. Noorhana Yahya received her M.Sc. and Ph.D. in Physics from Universiti Putra Malaysia in 1998 and 2002, respectively. She is presently a Professor in Fundamental and Applied Science Department of Universiti Teknologi PETRONAS. She is a physicist specializing in electromagnetics and nanotechnology. She heads the Advanced Electromagnetics Laboratory in Universiti Teknologi PETRONAS.

She managed to boasts of research grants worth over RM 70 million. In 2012, she leads a 12 million worth project funded by Long Term Research Grant Scheme (LRGS) secured from Malaysian Ministry of Higher Education (MOHE). This project was done with esteem collaboration from different public universities in Malaysia: Universiti Putra Malaysia (UPM), Universiti Sains Malaysia (USM), and Universiti Teknikal Malaysia Melaka (UTeM). The project was successfully completed in 2016, and to carry forward to the next level, strategic commercialization and integration with Internet of Things (IoT) elements are actively being conducted to ensure the project is in line with Fourth Industrial Revolution.

She has completed about 10 research projects, amassed 19 patents, and authored over 250 publications for the past 16 years. Apart from that, she also has been recognized within the nation and internationally. In 2010, due to her success in creating an electromagnetic transmitter for deep sea water exploration, she garnered a Gold Medal recognition the Science & Technology Competition held in Brussel, Belgium. She received the "Best Inventor Award" by the International Federation Inventor Association (IFIA) through International Technology Exhibition (ITEX) which was held in Kuala Lumpur. This award is a recognition for her remarkable creativeness and innovation in fertilizer research and is considered as the most prestigious award throughout her career as an academician and researcher. In addition, the Green Urea project bagged a gold medal in Nuremberg, Germany, as the "Best Green Invention" under the International Trade Fair Ideas-Inventions-New products (iENA) and four more from national-level research competitions.

Her most recent project is the RM 42.3 million "Game Changing Well Technology Using Electromagnetic." This project deals with the application of Nanotechnology and Electromagnetics to enhance oil recovery. The project collaborates with several national as well as international universities, which include Universiti Teknologi Malaysia (UTM) and Pembroke College, Cambridge UK.

Abbreviations

ACN	Acetonitrile
AI	Artificial intelligence
AN	Ammonium nitrate
ANOVA	Analysis of variance
B.C.E	Before common era
BFGS	Broyden–Fletcher–Goldfarb–Shanno
CAP	Common Agricultural Policy
CASTEP	Cambridge Serial Total Energy Package
CH_4	Methane
CO_2	Carbon dioxide
CPS	Cyber-physical systems
CRCGU	Controlled Release Coated Green Urea
CRF	Control release fertilizer
CT	Conventional tillage soil
DFT	Density functional theory
DOS	Disk operating systems
DMD	Dicyandiamide
DMPP	Dimethylphenylpiperazinium
EDX	Energy-dispersive X-ray spectroscopy
EEF	Enhanced efficiency fertilizer
EMF	Electromotive force
FCC	Face-centered cubic
FESEM	Field emission scanning electron microscopy
FTIR	Fourier-transform infrared spectroscopy
GHG	Greenhouse gas
GPS	Global Positioning System
HCls	Helmholtz coils
HYVs	High yielding varieties
IFA	International Fertilizer Industry Association
IoT	Internet of Things

IPCC	Intergovernmental Panel on Climate Change
IR 4.0	Industrial Revolution 4.0
IRRI	International Rice Research Institute
MFEs	Magnetic field effects
N	Nitrogen
N_2O	Nitrous oxide
NH_4	Ammonium
NH_4Cl	Ammonium chloride
NO_3	Nitrate
NT	No tillage soil
NUE	Nutrient use efficiency
Os	Osmium
PDOS	Partial density of state
PRB	Population Reference Bureau
RSM	Response Surface Methodology
SRF	Slow release fertilizer
TEM	Transmission electron microscopy
UN	United Nations
UNFCCC	United Nations Framework Convention on Climate Change
VSM	Value stream mapping
WFPS	Water-filled pores space
XRD	X-ray Powder Diffraction

List of Figures

List of Tables

Executive Summary

Climate change is an issue that will adversely affect human beings and other living species. Carbon dioxide (CO_2), methane (CH_4), and nitrous oxide (N_2O) are some of the greenhouse gases that would retain heat from the sun in the atmosphere, which leads to global warming. The global concern about deteriorating environment conditions necessitates the need for environmentally friendly technologies. Fertilizer production consumes approximately 1.2% of world's energy usage and is responsible for the same magnitude (1.2%) of the total greenhouse gases emission. The extensive use of conventional fertilizers, particularly the nitrogen-based fertilizer, in this case urea, had created environmental, economic, and social implications. Typical urea production requires high energy input, approximately 450°C and 250 bar. This motivates the research to have lower energy-intensive urea production. This game-changing technology, from a resource to a knowledge-intensive-based industry, investigates a new synthesis approach, employing electromagnetic induction and nanocatalyst at lower energy consumption. The clean and green method is for sustainable future, which might change the landscape of future chemical processes. However, this is made possible by the recent enhancement in nanotechnology where quantum mechanics is now well understood. New reactor designs are elaborated and discussed explicitly. Hematite and nickel oxide nanocatalysts were proposed for the green urea synthesis process, in the presence of static and oscillating magnetic fields. Strategies to increase single to triplet conversion rate are done for better understanding of the improved urea rate. The focus on scrutinizing greenhouse gas effect on the urea yield, in this case CO_2 flow rate, is deliberated explicitly. Coating techniques for slow release strategies are done to reduce the volatilization of ammonia and leaching effect, hence a complete solution of green technology. Agriculture 4.0 creates the new patterns and precision monitoring of crop rotation and livestock utilization, thus paving way for better crop yield. Development of advanced technologies for agriculture is an important task in the implementation of Agriculture 4.0, currently an inevitable trend of the socioeconomic development in the context of broader international integration for sustainable future.

Chapter 1
Urea Fertilizer: The Global Challenges and Their Impact to Our Sustainability

1.1 Introduction

Climate change is an issue that will adversely affect human being and other living species on earth. Carbon dioxide (CO_2) and other greenhouse gases such as methane (CH_4) and nitrous oxide (N_2O) retain heat from the sun in the atmosphere that leads to global warming. Svante August Arrhenius was the first scientist to address this issue back in 1896. His paper focused primarily toward defining the effect of carbon dioxide which he termed "carbonic acid" on universal cooling and where warming was measured as a consequence [1]. The study was done on two premises: (i) Heat suffers selective diffusion on its pathway through air and (ii) atmospheric gases absorb considerable quantities of heat. The temperature variation in the incident of growing CO_2 absorption in the atmosphere was estimated to be theoretically as high as 5 or 6 °C. It was also revealed that the temperature of the arctic regions would rise about 8 or 9 °C, if the concentration of CO_2 is amplified by 2.5 to 3 times of its existing value [2]. Years later, CO_2 concentration in the atmosphere has increased much more rapidly than he anticipated, but the earth has not warmed as he anticipated. Arrhenius calculations on the effect of carbon dioxide (CO_2) on global warming, as seen in Fig. 1.1, are regularly mentioned by the followers of anthropogenic, and the substantial or catastrophic warming would take place due to a rise in CO_2 [2].

Arrhenius studies on the CO_2 gas and its effect on global warming were forgotten for a very long time. It was later thought that the human influences were insignificant compared to the natural forces, such as volcanic effect, solar activity, and ocean circulation. For many years, scientists assumed that the CO_2 emission to the atmosphere will be absorbed by the ocean and hence contributes to the reduction of CO_2 pollution. It was found later that only 50% of the CO_2 pollution is absorbed by the ocean and land.

Climate change and global warming emerged as a major political issue in 1988 through Industrial Revolution which was addressed in Montreal Protocol [3]. It was

N. Yahya, *Green Urea*, Green Energy and Technology,
https://doi.org/10.1007/978-981-10-7578-0_1

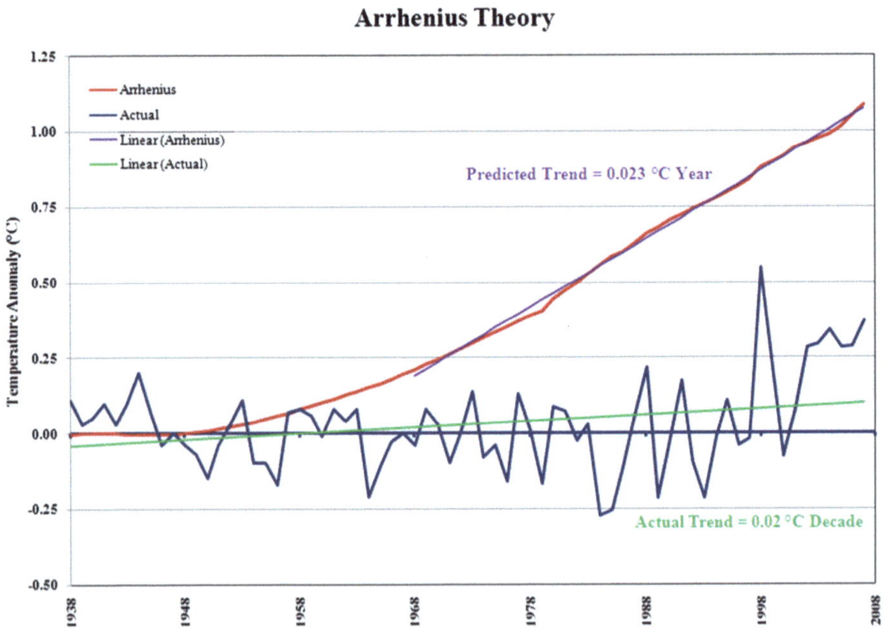

Fig. 1.1 Arrhenius plot to predict the global warming [2]

concluded that some of the greenhouse gases production is to be restricted initially and later banned ultimately due to the fact that it would cause the thinning of ozone layer, hence causing skin cancer to human. Later, in 1990, a broad consensus among scientists in the field that the possibility of global warming had to be taken seriously which was reported in the Intergovernmental Panel on Climate Change (IPCC) [4, 5, 6]. Only in 1997, most scientists considered that rising concentration of greenhouse gases in the atmosphere impacts the earth temperature and hence the change of weather patterns.

Kyoto Protocol was adopted in Kyoto, Japan, December 1997, which implemented the objective of United Nations Framework Convention on Climate Change (UNFCCC) to fight global warming by reducing greenhouse concentration to an acceptable level to prevent dangerous anthropogenic interference. A series of measurements from the top of Mauna Loa, the Hawaiian Volcano, demonstrated that CO_2 concentrations were rising steadily in the atmosphere, hence leading to series of meetings, mostly organized by various United Nations agencies. Scientists then began to realize that the increased concentration of CO_2 and other greenhouse gases would result in the climate change.

Today, global warming is due to modern living, transportation, fossil burning, agriculture, and other activities. It should be noted that CO_2, the currency of climate change, is affecting the well-being of mankind. One ton of CH_4 and N_2O is equivalent to 34 tons and 298 tons of CO_2 emission, respectively [7]. On top of this, N_2O has an atmospheric lifetime of 110 years and its removal process from the atmosphere depletes the ozone layer. Hence, an urgent attention should be called for the reduction of N_2O emission which comes mainly from the agriculture sector due to its alarming contribution in the global warming and the possible contributor to skin cancer [7] and potentially other diseases as well.

1.2 World Population

Connecting the global warming of the past and relating it to the future is important for human being and the entire ecosystem which are the center for future sustainability. The growth is also signifying a hope for a better tomorrow with well-educated people free from hunger and free from diseases who could give back to the mankind in improving global well-being [2]. Looking back, at 8000 BC, the world population is approximately 5 million. This was at the dawn of agriculture where man starts to plant crops. In 1800, there was a sharp increase in the world population during the birth of the First Industrial Revolution, where the world population was about 1 billion.

In 1930, the second billion populations were achieved during the Third Industrial Revolution. In the year of 2017, where the Fourth Industrial Revolution takes place, the population had further increased up to 7.5 billion and is expected to increase up to 9.8 billion in 2050 [8]. It is obvious that the population grew slowly for the past thousands of years and only in the recent centuries that we have observed the rapid population growth, approximately tripled between 1900s and 2000s [1].

Scrutinizing the trend, the world population has been accelerating for a long time. Nevertheless, the accelerated pattern has been diminished as the rate of population growth has been decreased. The growth rate has dropped to almost half from its peak at 2.1% in 1962, as seen in Fig. 1.2. Hence, it shows the end of a long period of accelerated growth [2]. The distribution of the population according to the age is the one and only aspect that requires attention especially on how the young population will evolve. An aging population is expected in the next decade.

It should be noted by 2024 that India will overtake China population as the world's most populated country, and Nigeria has the world's fastest growing population. It was estimated that by 2050, India will be the most populated country, followed by China and Nigeria. The projection made by Population Reference Bureau (PRB) has indicated that by 2050 Africa's population will be more than double, up to 2.6 billion. This accounts for an approximately 57% of the global population.

The steady rise of population in 2050 calls for an urgent attention for food security. In addition, there might be potential increase in hunger and decrease in arable land. N_2O emission has increased mainly due to an increased use of mineral fertilization. The change in lifestyle diet has increased the demand for meat, which

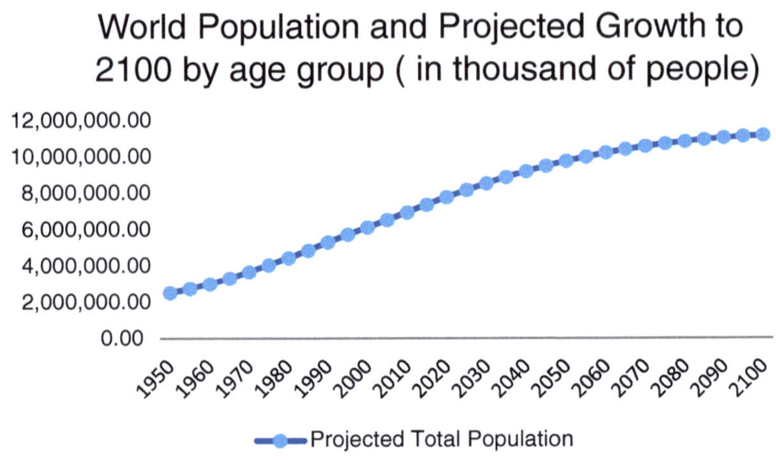

Fig. 1.2 World population and projected growth to 2100 by age growth [2]

also contributed to the need for increased agriculture production rate. This had resulted in some major impact in agriculture practices in adapting the global warming effect and hence adapting to climate change.

1.3 The World Fertilizer Market

Fertilizer plays an important role in ensuring sustainable crop production and global food security world widely. Improved forecast in the fertilizer demand is needed, to ensure that the supply for food of an increase world population is successfully met. The main concern of fertilizer utilization is either it is under or inefficiently applied, which causes poor crop yield or over application that consequently affects the environment. According to International Fertilizer Industry Association (IFIA), more than 70% of fertilizer consumption is dominated by China, India, and USA. China is the main producer of nitrogen- and phosphate-based fertilizer, while the leading producer of potash fertilizers is Canada, and then followed by Russia [9].

Plants need 27% of nitrogen (N) in the form of ammonium (NH_4) and nitrate (NO_3) in a ratio of 1:1. Nitrogen fertilizers in its raw constituents are largely in the form of natural gas and methane, which is known as the greenhouse gas. Ammonia is used as the feedstock to produce fertilizers such as urea, ammonium nitrate, or ammonium phosphate [8, 9, 10, 11].

By 2020, the shortage of nitrogen-based fertilizer is foreseen especially for the three largest producing countries (China, Indonesia, and USA) and this is based on the pattern of production and consumption of the nitrogenous fertilizer according to these countries. This may be due to many factors such as rising population, changing diets, better nutrient balance, weather patterns, crops and fertilizers prices.

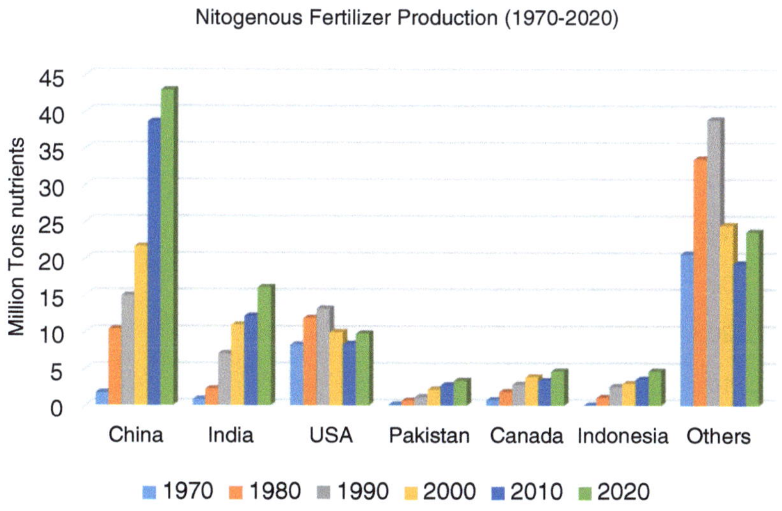

Fig. 1.3 Nitrogen fertilizer production per country (1970–2020) *Source* IFA (International fertilizer industry association) [17]

It is estimated that from 2010 to 2020 China would increase its consumption rate about 75% of nitrogen fertilizer while production increases are only about 11%. The same scenario is expected for India and USA by 2020.

From 1970 to 2010, in every ten years of interval, Pakistan and Indonesia showed higher production rate of nitrogen fertilizers with lower consumption. However, both countries are estimated to have higher consumption rate in 2020. Meanwhile Canada recorded low consumption of nitrogen fertilizer with higher production throughout the interval period and expected to continue with this trend till 2020. The consumption is due to the main agricultural activity in each country. It is noted in Fig. 1.3 that China, India, and USA are the world's major consumers of nitrogen [12, 13, 14, 15, 16]. The world production and consumption from 1970 to 2020 are represented in Figs. 1.3 and 1.4.

1.4 Optimization of the Nitrogenous Fertilizer of the World Production and Consumption

The Response Surface Methodology (RSM) is used to evaluate the nitrogenous world fertilizer production and consumption data from 1970 to 2020. The purpose of optimization is to know the effect of the world major producing countries of the nitrogenous fertilizer to the world market. The design is developed to study the relationship between the variable (year) and the response (production and consumption).

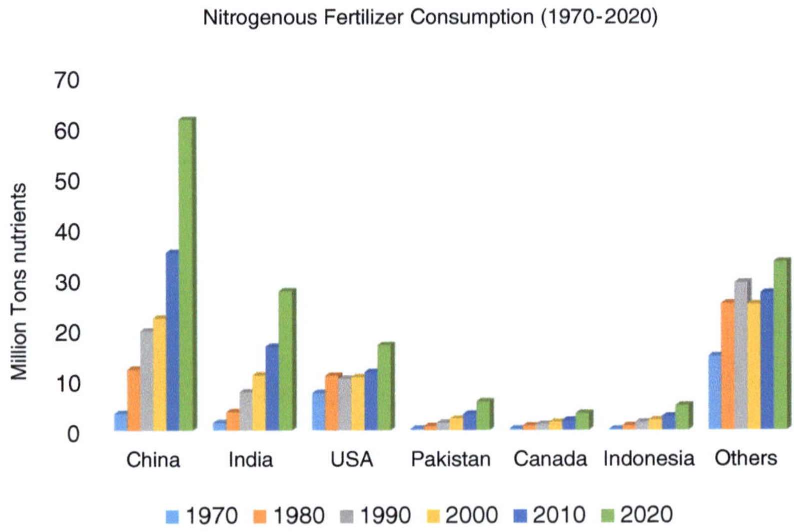

Fig. 1.4 Nitrogen fertilizer consumption per country (2070–2020) *Source* IFA (International fertilizer industry association) [17]

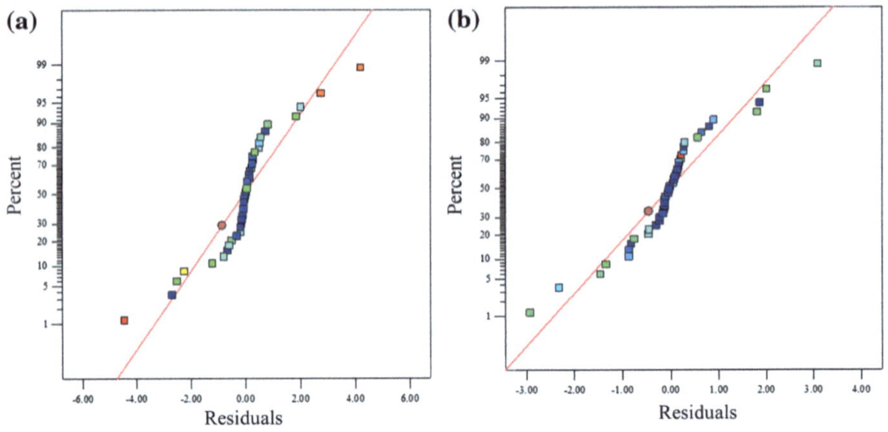

Fig. 1.5 Normality plot of residual for **a** production **b** consumptions

Analysis of variance (ANOVA) is used to test whether the variable (year) mean is significantly affecting the response values (production and consumption). The F-values were calculated to measure the size of the effects (Fig. 1.5).

Figure 1.6a, b shows the graph plots of the normality test of the residuals for the nitrogenous fertilizers production and consumption, respectively. It is observed that in the both graphs, the residuals lie on the straight line which indicates that the distribution of residuals for both production and consumption is normal.

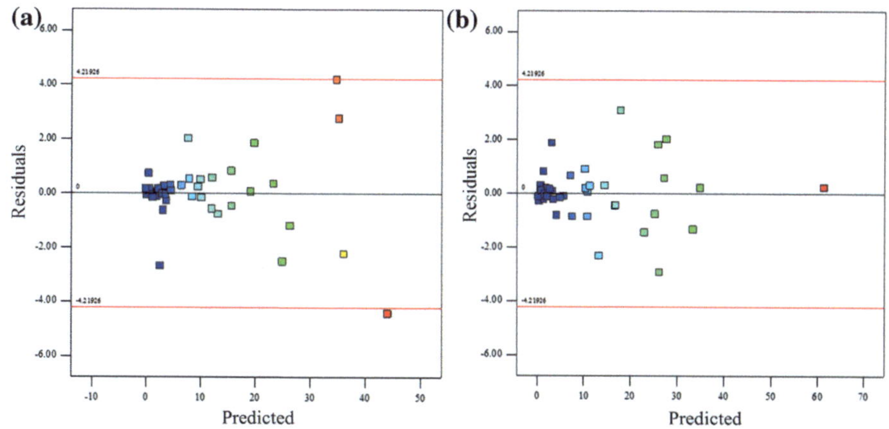

Fig. 1.6 Plots of predicted versus residual for nitrogenous fertilizer from 1970 to 2020 for its **a** production **b** consumption

Residuals versus predicted response for the nitrogenous fertilizers production and consumption, respectively, are depicted in Fig. 1.6a, b. The graphs show that most of the points lie within the area of ±3.0. This indicates that the assumption of constant variance is confirmed, and thus, the suggested model is suitable.

Tables 1.1 and 1.2 show the ANOVA results for the nitrogenous fertilizers of the production and consumption, respectively. The objective of the world nitrogenous fertilizer optimization of the production and consumption is to visualize the supply and the future demand of nitrogenous fertilizer of the whole world, if it can meet the need of the global fertilizer hence agriculture industry.

The two F-values of the production and consumption models are 37.76 and 283.40, respectively, which implies that the two models are significant. In these two models, there is only a 0.01% chance of getting a large F-value which is due to the noise. The probability (P-value) for the production model terms A, B, AB, A^2B, and A^3B and for the consumption model terms A, B, AB, A^2, A^2B, A^3, and A^3B was significant because the values are less than 0.05.

Table 1.1 Nitrogenous fertilizer production model data from 1970 to 2020

Source	Sum of squares	df	Mean square	F-Value	p-value Prob > F
Model	5349.39	28	191.05	37.76	<0.0001
A-Year	*394.56*	*1*	*394.56*	*77.98*	*<0.0001*
B-Country	*3577.64*	*6*	*596.27*	*117.84*	*<0.0001*
AB	*1098.10*	*6*	*183.02*	*36.17*	*<0.0001*
A^2B	*89.18*	*6*	*14.86*	*2.94*	*0.0488*
A^3	*29.17*	*1*	*29.17*	*5.76*	*0.0320*
A^3B	*147.35*	*6*	*24.56*	*4.85*	*0.0082*

Table 1.2 Nitrogenous fertilizer consumption model data from 1970 to 2020

Source	Sum of squares	df	Mean square	F-Value	p-value Prob > F
Model	6832.89	28	244.03	283.40	<0.0001
A-Year	*1182.50*	*1*	*1182.50*	*1373.26*	*<0.0001*
B-Country	*4015.04*	*6*	*669.17*	*777.12*	*<0.0001*
AB	*1302.47*	*6*	*217.08*	*252.10*	*<0.0001*
A^2	*49.73*	*1*	*49.73*	*57.75*	*<0.0001*
A^2B	*138.83*	*6*	*23.14*	*26.87*	*<0.0001*
A^3	*72.96*	*1*	*72.96*	*84.73*	*<0.0001*
A^3B	*70.48*	*6*	*11.75*	*13.64*	*<0.0001*

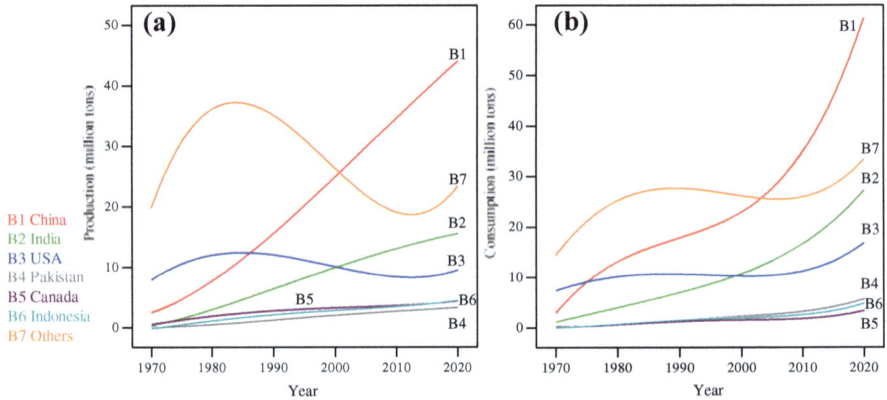

Fig. 1.7 **a** Production and **b** consumption of the nitrogenous fertilizer versus the year

Figure 1.7a, b shows the consumption and production of the nitrogenous fertilizer, respectively, for different countries at different years. From the graphs, it is clearly shown that the supply and demand of the nitrogenous fertilizer are increasing by year from 1970 to 2020. Regarding the consumption of the nitrogenous fertilizer, China consumption increases gradually from 1970 to 2010 then it starts to increase sharply on the following year onwards.

It is clearly shown in the graph that China has exhibited the highest growth rate in both the production and consumption of the nitrogenous fertilizer. This is followed by India that shows a similar trend. It is noted that the production of nitrogenous fertilizer of the USA is nearly constant. There is a slight fall of the production in 1990 and 2010. The consumption of this type of fertilizer is also similar with a slight increase in the demand between the years of 2010 to 2020. The other three countries, Pakistan, Canada, and Indonesia, have similar trend of the production and consumption. It is noted that for the other countries there is a sharp increase in the production trend, from the year of 1970 to 1985 before a drastic decrease in 1985 to 2010. The trend of consumption for the other countries is similar to that of the trend of consumption in the USA. It is obviously seen that China only became the major nitrogenous fertilizer production in 1987. It has consumed more than the amount that has been produced in the year of 1970.

The models that are used to predict the production and consumption of the nitrogenous fertilizer are represented by the Eqs. 1.1 and 1.2 in terms of actual factors, respectively.

$$Production = a - by + cy^2 - dy^3 \tag{1.1}$$

$$Consumption = e - fy + gy^2 - hy^3 \tag{1.2}$$

In general (Table 1.3), the consumption and production of each country can be predicted by the Eqs. 1.1 and 1.2,

where y is the year and *a*, *b*, *c*, *d*, *e*, *f*, *g*, and *h* are constants varying with countries as shown in Table 1.3.

The above equations show that the trend is going to increase year by year for the fertilizer production and consumption in the whole world.

As clearly seen in Fig. 1.8, as predicted by RSM model, in the year of 2050 the consumption of the nitrogenous fertilizer in China is 3 times higher as compared to its consumptions in 2030. As compared to the three countries in the year of 2050, the nitrogenous fertilizer consumptions in China are greater by 2.5 times, as compared to India, and 3 times greater as compared to the USA. However, it is also predicted that, in the year of 2050, the overall nitrogenous fertilizer consumptions in China will be 5 times higher compared to its production.

As in India in the year of 2050, the consumption of the nitrogenous fertilizer is predicted 2.7 times higher as compared to its consumptions in 2030. However, its overall consumption is predicted to be 6.2 times higher compared to its production, in the year of 2050.

While in the USA, in the year of 2050, the consumption of the nitrogenous fertilizer is predicted 3.44 times higher as compared to its consumptions in 2030. It is also shown in Fig. 1.8 that its overall consumptions are predicted to be 1.8 times higher compared to its production, in the year of 2050.

Table 1.3 Constant (a, b, c, d, e, f, g, and h) varying with countries for the evaluation of the production and consumption of nitrogenous fertilizer, from the year of 1990 to 2020

Country	a	b	c	d	e	f	g	h
China	16962692.20	33299.32	24.51	0.01	53548699.34	111684.86	87.24	0.03
India	16207324.56	32172.97	23.95	0.01	60292369.34	121846.78	92.34	0.03
USA	12503391.52	26609.21	21.16	0.01	59224766.39	120252.79	91.55	0.03
Pakistan	15754034.23	31489.79	23.60	0.01	61764608.40	124074.21	93.47	0.03
Canada	15183801.34	30634.33	23.18	0.01	61767573.28	124080.82	93.47	0.03
Indonesia	15265979.20	30757.28	23.24	0.01	61593221.11	123817.60	93.34	0.03
Others	2850803.05	12115.13	13.91	0.01	55179858.22	114185.47	88.51	0.03

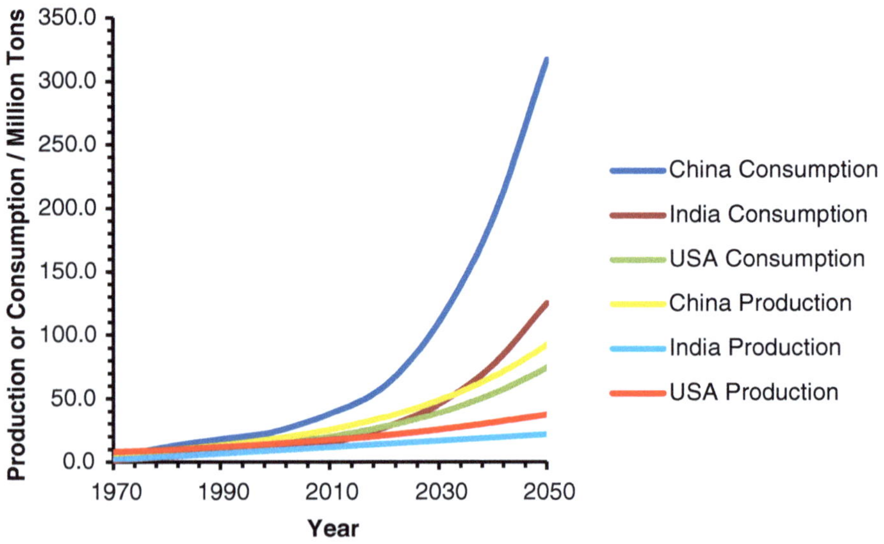

Fig. 1.8 Production and consumption of the nitrogenous fertilizer versus the year of China, India, and the USA

It is obviously shown that the RSM method is a very powerful tool for the prediction of the need, of the population in regards to its food consumption by the predictive model.

1.5 Global Warming Impact of Nitrogenous Fertilizers

Agriculture yield can be increased by using fertilizers, minerals, pesticides, and herbicides. The urea, nitrogen-based fertilizer production, is an energy-intensive industry which uses methane and carbon dioxide as the feedstock. Greenhouse gas emissions from the application of fertilizer through volatilization can be controlled. A proper understanding of the interplay of the environment, socioeconomic, and technical aspects of the technology is crucial for a successful generation of the new agricultural technologies that affect our livelihood.

The imperative studies can be carried out, in the dimensions of economics and sociological development, to complement the research work outcomes by the scientist. Besides the environment dimension that has been mentioned earlier, the sociological implication of the technology among farmers and smallholders, their current farming practices, institutions governing, diffusion of agricultural technologies, and the market for agricultural technologies must be embedded and integrated. This is important as quite often the success or failure of the new agricultural technologies hinges critically on the social and economic dimensions apart from the technical viability of the technology.

The production of nitrogen fertilizers has great impact on the climate policies. The world fertilizer industry has decided to make a future concerned with energy and climate road map up to 2050 [18, 19]. The world fertilizers companies have supported in the preparation of this road map. This road map focuses on two of the most important N-fertilizers, ammonium nitrate (AN) and urea, given by the energy-intensive production of nitrogen (N) fertilizers.

The challenges are associated with the growing threats to the global warming potentials that can only be resolved if the major economics country contributor, such as the US and China, give their full attention on the future sustainability. Slow release properties utilizing the coating technology are expected to be emphasized by reducing the volatilization and denitrification effect.

Nitrogen cycles as shown in Fig. 1.9 consists of three major parts, namely, nitrogen fixation, nitrification, and denitrification. 78% of air comprises of nitrogen. However, neither plants nor animals are able to utilize the nitrogen from the environment. Living organism depends on nitrogen fixation which is an anaerobic process that converts atmospheric nitrogen to ammonia by bacteria [20]. This ammonia is transferred from the bacteria to the plant for chlorophyll, nucleic acids,

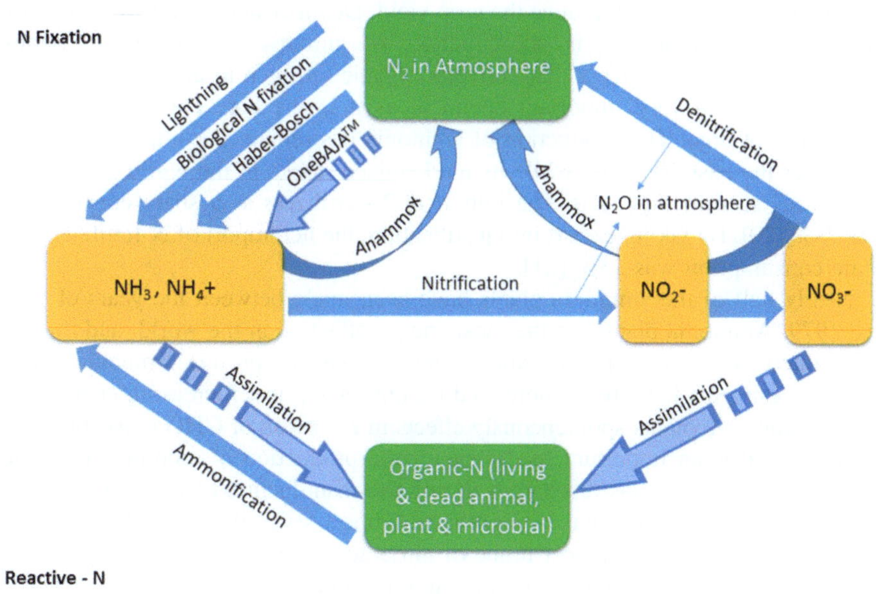

Fig. 1.9 Diagram of the nitrogen cycle, showing transformations from unreactive N_2 in the atmosphere to the various forms of reactive-N and back to N_2, including the processes of N fixation (lightning, biological N fixation, and the synthetic Haber–Bosch process), nitrification, assimilation into plant, animal, and microbial biomass, ammonification, and denitrification. Atmospheric nitrous oxide (N_2O) is produced as a by-product of microbial nitrification and denitrification [20]

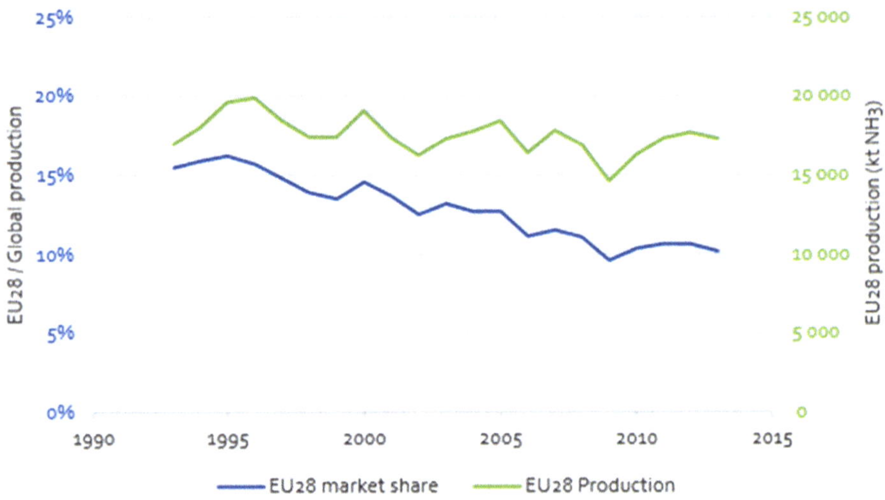

Fig. 1.10 Ammonia production in EU28 (in kt NH_3) and the EU28 market share in the m world [22]

and enzymes synthesis. Nitrogen is an essential nutrient for all plants, and deficiency in this nutrient will reduce the crop yield. Denitrification is a reverse process of nitrogen fixation. It is a reduction of nitrates to nitrogen gaseous. Ammonification is a process that converts organic nitrogen to ammonia.

Ammonia is a very important chemical as it is the core component for the production of urea. The production of ammonia has remained comparatively constant over the past 20 years, while its marketplace stocks in the world are steadily reduced as seen in Fig. 1.10. The import of N-fertilizers as a share consumption was 28% [20]. By taking export into justification, the net import of N-fertilizers as a share consumption was 13% [21].

Nearly half of the ammonia plants have been made between the years of 1970 and 1979. Ammonia plants are the most energy effective in the world, and this can be credited to the struggles to expand and upgrade the present ammonia plants. These investments have been motivated by high gas price. Increasing energy efficacy of ammonia plants spontaneously effects in reduction of GHG emissions [23].

Nitric acid is an important feedstock for the production of ammonium nitrate. The main greenhouse gas discharges during the manufacture of nitric acid are the emissions of N_2O. This has a significant impact over the past decade (87% decreases in N_2O emissions per tonne of nitric acid in 11 years, matching to 17% per year). In the manufacturing of ammonium nitrate, the remaining N_2O released from the production of nitric acid is almost 5 times lesser than the CO_2 emissions from the manufacturing of ammonia [24].

Fertilizer manufacturers will contribute toward the discharge of chemicals which leads to the environmental hazard. During the application ("use"), fertilizers cause

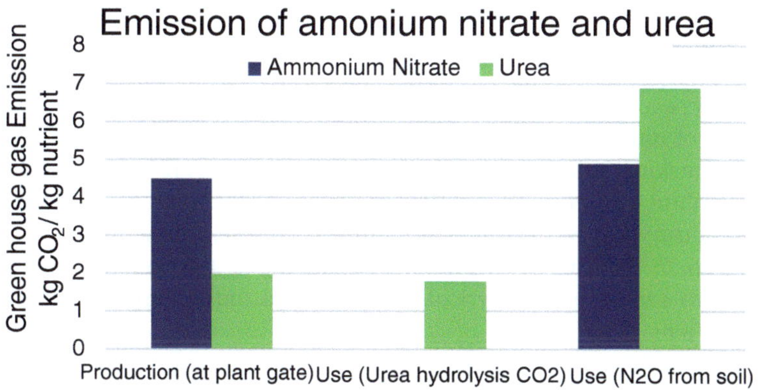

Fig. 1.11 Fertilizers carbon footprint reference values for mineral fertilizer production and use, 2011–14 (not covering all emission sources) [22]

further discharges. To illustrate this, a few important release sources during the "use phase" are shown in Fig. 1.11.

Figure 1.11 displays that the CO_2 "captured" in the urea in its built-up is released during its consumption. The leading releases during the use of fertilizers are yet N_2O. N_2O is formed by bacteria in all soils. In agriculture, N_2O is released mainly from fertilized soils and animal wastes whereas nitrogen (N) is freely obtainable [24].

Referring to Fig. 1.12, the N_2O released, when making nitric acid was 0.23 CO_2 eq/tonne nitric acid in 2013, although there are plans to further decrease the release rate to ~0.09 ton CO_2 eq/tonne nitric acid. Throughout the production of nitric acid, heat is made (6.3 GJ/tonne 100% nitric acid), which is partially used.

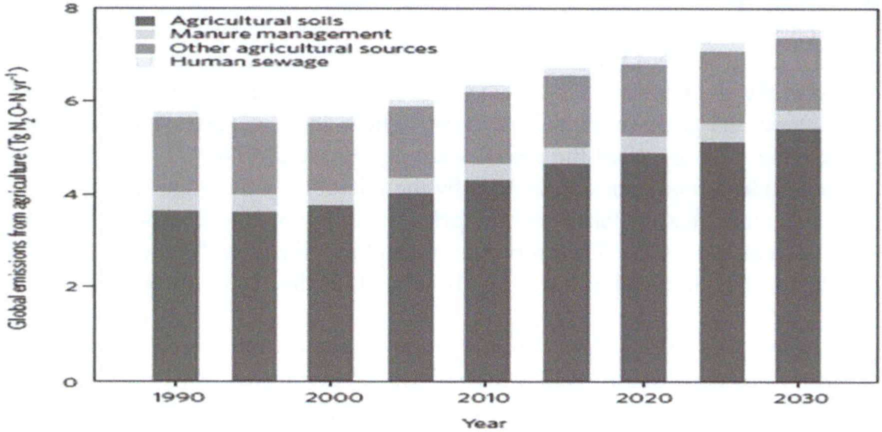

Fig. 1.12 Global N_2O emissions from agriculture between 1990 and 2030 [24]

Distributing the use of this heat in other places has lowered the emissions of the greenhouse gas.

The production is the key building block of ammonia which needs by outlying most energy, and ammonia plants are the utmost energy capable in the world. The production of nitric acid traditionally leads to a significant amount of N_2O emissions, thus making it the utmost greenhouse gas emissions [25, 26]. These releases have been abridged with 87% between the year of 2002 and 2013. Therefore, the fertilizer industry can be measured as a prime candidate in limiting greenhouse gas releases from the built-up of fertilizers. Apart from that, the manufacturing fertilizer subsidizes to switch on the agricultural yield and thus promotes safety of food supply and jobs.

1.6 The Emissions of NO_2 and N_2O from Soils and Its Model

Nitrous oxide is an important greenhouse gas, due to its high global warming potential. Agricultural soils are the main source of N_2O emission to our atmosphere. The use of N-fertilizer is important to achieve a desired yield, but N can be lost to the atmosphere as the emission of N_2O, thus enhancing the greenhouse effect. The annual increase rate of N_2O varies from 0.2% to 0.3%. Thus, any strategy in decreasing the greenhouse gases (GHG) concentration in our atmosphere has to focus on the agricultural sector, because this is the main source of N_2O.

Our atmosphere is composed by 78% of nitrogen (N_2) and 28% of oxygen (O) [27]. One-third of the cosmic radiation that intercepts the outer atmosphere is bounced back to space. The remaining two-thirds pass through the atmosphere and are absorbed by the earth's surface and its atmosphere. Then, the earth surface emits it back in the form of infrared rays. Mostly this energy is absorbed by the atmosphere and then reflected back to the earth's surface. This process is known as the greenhouse effect [27].

CO_2, CH_4, and N_2O are considered as the greenhouse gases (GHG) that are related to the greenhouse effect. Each GHG absorbs the infrared radiation from our atmosphere and then re-emits it as heat, which then increases the atmosphere temperature. This is called the global warming potential (GWP).

The synthesis and emission of N_2O by the microbial activity are induced by the soil temperature, texture, structure and pH, availability of N, decomposable organic material, and water [28]. The crop rotation, soil mobilization, N sources, and time and depth of N application have greatly influenced the N_2O emission by the soil [29].

The emissions of N_2O from denitrification are mainly influenced by the amount of water-filled pores space (WFPS). Despite the importance of N to nitrification and denitrification reactions, these processes are also influenced by soil characteristics, such as aeration, compaction, temperature, moisture, pH, organic matter, available N, C/N ratio, texture, and also by soil management and crop rotation [30].

It is found by Akiyama et al. that during the peak emission of N_2O it is observed a decrease in the amount of NH_4^+ and increase of NO in the soil [31]. It is observed that the emission of N_2O is constant when the nitrification is kept constant and the concentration of NO concentration is minimal [32].

The temperature of soil and its moisture strongly influence the emission of NO and its diffusion to our atmosphere [33]. As the temperature increases, the rate of nitrogen conversion is increased [34]. It is also found that there is a close relationship between seasonal variation of NO emission, soil, and air temperatures [35]. Also found that increasing soil temperatures stimulates soil respiration from the microbial activity and thus increasing anaerobic sites for denitrification process.

It is also found that an increase in the soil moisture contributes to the greater emission of N_2O emission [36]. The nitrification and denitrification processes are mainly influenced by moisture [37]. However, the NO production decreases at an extremely moisture condition [38]. This is because the microbial activity is inhibited at very high level of soil moisture. However, it is found that at very high water content in the soil decreases NO emissions [39].

When the soil aeration is lower, N_2 is the main nitrogen gas emitted to the atmosphere [38]. Soil aeration has a close relationship with soil moisture. The greater the WFPS, the smaller the number of soil pores filled with air, enhancing the production of N_2O by denitrification. The amount of WFPS has a great correlated with the carbon, nitrification potential, and mineralization potential, and thus showing the importance of moisture to microbial activities in order to produce N_2O in the soil [38]. The general assumption as stated by Denmead et al. that the formation of NO in soils is enhanced when the WFPS reaches approximately 80% [39].

It is also found that when the WFPS is higher than 70%, the denitrification is the dominant process in producing N_2O, but when the WFPS decreases to 60%, it is found that the process of nitrification is activated. It is important to highlight here that the occurrence of dominant process as denitrification will not guarantee the high emission of NO [39].

N_2O emissions may be influenced by the relief position, since it affects the moisture dispersion, C organic content, and soil fertility. The low relief areas tend to be more humid and to have greater organic matter content as compared to higher lands. It also found that the emission of NO is also influenced by the respiration activity of microorganism and its oxygen consumption [38].

The emission of N_2O is influenced by the type of soil [40]. It is also found that clayey soils tend to have greater NO emissions than sandy soils. Chen et al. (2008) found that soils of fine texture and immobilization before seeding phase contribute to the higher emission of N_2O.

Denitrification processes require a variety of metal cofactors, including Mo, Fe, Cu, and Zn. The absolute requirement of N_2O reductase is the metal cofactor of Cu which is so critical in the process of denitrification. This is an important aspect, with agricultural and environmental implications [41].

The production of N_2O by denitrification is favored by high concentrations of NO and low pH in incubated soils [45]. For low pH values, the emission of NO would be greater, because of a small amount of this gas would be reduced to N [43].

An increase in one unit of pH may then decrease 0.2 units in the molar fraction of the NO emitted [40]. Thus, the continuing acidification of agricultural soils through excessive use of nitrogen fertilizers could drastically enhance N_2O emissions, while the careful adjustment of soil pH by liming would reduce emissions [41]. The nitrification process which is aerobic tends to reduce the soil pH, while the denitrification which is anaerobic can increase the soil pH [32].

Brentrup et al. (2000) also stated that increasing the soil organic C content can increase the NO production. Furthermore, Ciampitti et al. (2008) found that the soil moisture and soluble C content show a significant correlation with N_2O emissions, during residue decomposition.

The concentration of C influences the nitrification and denitrification reactions [44]. This is because it stimulates the microbial growth and its activity and also provides the organic carbon needed by soil denitrifies [45]. It is found that the microbial growth and its metabolism increase the consumption of O_2 and generate an anaerobic condition that is necessary for denitrification [46].

In dry and well-aerated soils, nitrification is the dominant process and the gas diffusion is greater, allowing a greater emission of the more oxidized form of NO. In moist soils, the rate of gas diffusion and aeration is smaller, and greater amount of NO would react before being released to the atmosphere. It is also found that under supersaturation conditions (anaerobic soil), most of the N_2O is reduced to N_2, before being released to the atmosphere [46]. The consumption of N_2O also depends on the soil properties, water content, soil temperature, pH, and availability of organic C and N. The longer NO remains in the soil, more O will be used as electron acceptor, and thus the emissions of N_2O will be higher.

The climatic conditions, soil management practice and the metabolic activity of the microbial, influenced the production of N_2O. The most contributing factors toward the emission of N_2O are the soil tillage, recycling N from crop residues, and the application of N-fertilizers.

However, there are considerable shortcomings in the current state of knowledge on how dynamic factors can influence the denitrification process. It is stated that much of this is due to the high spatial [47] and temporal variability of denitrification rates in the field and the failure to include functional differences in the microbial composition, in models that estimate N_2O emissions by Cavigelli and Robertson [48].

The higher soil moisture is due to the crop residue in no-tillage soil (NT), thus increased the microbial activity near the soil surface, and creating anaerobic microsites due to higher consumption of O_2 [49]. N_2O emissions in a crop rotation system, in China, showed that the incorporation of maize and wheat straw significantly increased the soil temperature, due to their heat-retaining property. This higher temperature could have stimulated the enzymatic activity of nitrifiers and denitrifiers, enhancing the microbial NO production. In the conventional tillage soil (CT) system, such mechanism would be dissipated by the tillage in the upper soil layer, which increases the O_2 concentration in soil and consequently decreases the NO emission. Also found that the greater mineralization rate of crop residue and organic matter is promoted by soil tillage [50]. The soil tillage and plant residue have increased the emissions of NO and CO_2 [51].

The area with recent adoption of NT which can be up to 10 years old showed higher NO emissions as compared to CT [52]. In much older of NT which is 20 years old, it showed the lower emissions of NO as compared to CT. It is also observed that the emission of N_2O is similar in dry area of both NT and CT.

It is also showed that NT in humid areas increases N_2O emissions in early years and then reduces them, in comparison to CT [53]. The emission of NO for the 30 years old of NT areas is lower as compared to CT [54]. The main reason for the low rate of N_2O emission in the areas with early NT system is that the soil moisture has been changed, favoring the denitrification process, and thus causing N loss, inducing N deficiency in plants, and finally reducing the crop production [52].

Before 1950, less than 50% of NO emissions were caused directly or indirectly by the use of N-fertilizers in agriculture, being most of the emissions associated to animal breeding. As a result of the increase in food production and the consequent use of N-fertilizers, two-thirds of the N_2O emissions were related to crop production, in 1996 [54].

The emission of N_2O by the nitrification and denitrification processes depends on the content of N in the soil [31]. It is also found that the rate of NO emissions is related to the N-fertilizer applied in the soil [29]. The use of N-fertilizer influenced the amount of NH_4^+ or NO in the soil. The greater the amount of N-NH_3^- in the fertilizer, the nitrification process will be greater [55]. The use of N-fertilization at higher plant biomass production has created more residues from the crop in the soil, which then increases the emission of NO [56].

A mathematical relationship between accumulated NO emission and amount of N applied as fertilizer is not well defined. It is stated that the relation is linear and can be used to estimate the emission of N_2O in the soil [35, 40, 57]. The emission of N_2O is exponentially related to the N-fertilization rates [58]. The emission of N_2O induced by the ammonium nitrate fertilizer is higher as compared to urea fertilizer [29]. The fertilizer application depth also influences the NO emission [59, 60]. It is also found that the application of fertilizer during dry weather resulted in the lower emissions of NO, as compared to moist conditions [35].

The slow-releasing fertilizers reduced the emission of NO by N-fertilizers. It is also stated that the application of urea formaldehyde and urea with nitrification inhibitors (hydroquinone and dicyandiamide) reduced NO emissions [60]. The N-fertilizers stabilized by inhibitors are classified as slow-acting nitrogen.

The International Panel on Climate Change (IPCC) suggests a methodology to estimate soil NO emissions, according to mineral and organic fertilizers, and also for the mineralization of N from crop residues added to soil. The emission factor is used to determine the life cycle of agricultural crops [37]. The high emission factor value is reported by Denmead [38], and it is shown to be related to the high concentration of organic carbon and frequent rain, which then contributed to pore saturation and N_2O emission. In managing the N-fertilizer application, the emission of NO is reduced when N-fertilizers are applied in an appropriate depth, instead of surface application. Splitting the rate of N-fertilizer recommended to the crop and applying the fertilizers to the inner soil layers can be used to reduce NO emissions.

1.7 Summary

World population will increase and hence would result in a decrease in arable land. Based on the potential increase in world population, it is inherent that efficiency in fertilizer must first be considered to avoid starvation. In addition, to ensure the stability of the ecosystem, green technology needs to be addressed to ensure clean and green environment which must be given the priority. The green fertilizer is proposed utilizing the nanotechnology. Global potential nitrogenous fertilizer production and consumption indicated that China will be the highest consumed nation and the production rise is consistent. The relationship between production and consumption of nitrogenous-based fertilizer was evaluated to ensure that the world need for food is met. The agricultural soils are the main source of N_2O emission to our atmosphere. The use of N-fertilizer is important to achieve a desired yield, but N can be lost to the atmosphere as the emission of N_2O, thus enhancing the greenhouse effect. The sustainability in agricultural can be enhanced by improving the efficiency of the fertilizer used and minimizing the impact of soil management systems. It is important to determine the impact of the agricultural inputs to the biological process and its ecosystem.

References

1. Arrhenius, S. (1986). On the influence of carbonic acid in the air upon the temperature of the ground. *Philosophical Magazine and Journal of Science, 41*(270), 237–279.
2. Arrhenius, S. Ueber den Einfluss des atmosphärischen Kohlensäuregehalts auf die Temperatur der Erdoberfläche.
3. U. Nations. (1998). Kyoto protocol to the United Nations framework kyoto protocol to the United Nations framework. *Review of European Community and International Environmental Law, 7,* 214–217.
4. Panel, I., & Climate, O. N. (1997). Intergovernmental Panel on Climate Change Ipcc/Oecd/Iea Programme on National Greenhouse Gas Inventories Expert Group Meeting on Methods for the Assessment of Inventory Quality (co-sponsored by CKO/CCB), Bilthoven, The Netherlands 5–7 November 1997.
5. IPCC. (2007). Climate change 2007: the physical science basis. *Intergovernmental Panel on Climate Change, 446*(7137), 727–728.
6. IPCC. (2014). Climate Change 2014: Synthesis Report. Contribution of Working Groups I, II and III to the Fifth Assessment Report of the Intergovernmental Panel on Climate Change.
7. Ayoub, A. T. (1999). Fertilizers and the environment. *Nutrient Cycling in Agroecosystems, 55* (2), 117–121.
8. United Nations, Department of Economic and Social Affairs, Population Division. (2017). *World Population Prospects: The 2017 Revision*. New York: United Nations.

9. Potash, USGS Mineral Commodity Summary 2012.
10. Stehfest, E., & Bouwman, L. (2006). N2O and NO emission from agricultural fields and soils under natural vegetation: Summarizing available measurement data and modeling of global annual emissions. *Nutrient Cycling in Agroecosystems, 74*(3), 207–228.
11. Steffan, J. J., Brevik, E. C., Burgess, L. C., & Cerdà, A. (2017). The effect of soil on human health: An overview. *European Journal of Soil Science*.
12. Pinder, R. W., Davidson, E. A., Goodale, C. L., Greaver, T. L., Herrick, J. D., & Liu, L. (2012). Climate change impacts of US reactive nitrogen. *Proceedings of the National Academy of Sciences of the United States of America, 109*(20), 7671–7675.
13. Buyanovsky, G. A., & Wagner, G. H. (1998). Changing role of cultivated land in the global carbon cycle. *Biology and Fertility of Soils, 27*(3), 242–245.
14. Denman, K., Ciais, P., Cox, P., Dickinson, R., Hauglustaine, D., Heinze, C., et al. (2007). Couplings Between Changes in the Climate System and Biogeochemistry (p. 119).
15. Wofsy, S. C., Zhang, X., Qin, D., Manning, M., Chen, Z., Marquis, M., & Averyt, K. B. (2007) Couplings between changes in the climate system and biogeochemistry. *Climate Change 2007—The Physical Science Basis, 21*(7), 499–587.
16. Fertilizers Europe. (2013). EU fertilizer market. Key graphs.
17. E. Fertilizers. (2013). Closing the loop, pp. 1–7.
18. Galloway, J. N., Aber, J. D., Erisman, J. W., Seitzinger, S. P., Howarth, R. W., Cowling, E. B., et al. (2003). The nitrogen cascade. *BioScience, 53*(4), 341.
19. Cole, C. V., Duxbury, J., Freney, J., Heinemeyer, O., Minami, K., Mosier, A., et al. (1997). Global estimates of potential mitigation of greenhouse gas emissions by agriculture. *Nutrient Cycling in Agroecosystems, 49*, 221–228.
20. Davidson, E. A., Nifong, R. L., Ferguson, R. B., Palm, C., Osmond, D. L., & Baron, J. S. (2016). Nutrients in the nexus. *Journal of Environmental Studies and Sciences, 6*(1), 25–38.
21. Erisman, J. W., Sutton, M. A., Galloway, J., Klimont, Z., & Winiwarter, W. (2008). How a century of ammonia synthesis changed the world. *Nature Geoscience, 1*(10), 636–639.
22. Stork, M., & Bourgault, C. (2015). Fertilizers and Climate Change—Looking to 2050 (pp. 1–24).
23. Nash, M., & Neale, B. (2013). Fertilizer Europe (Vol. 23, no. 11, p. 13).
24. Reay, D. S., Davidson, E. A., Smith, K. A., Smith, P., Melillo, J. M., Dentener, F., et al. (2012). Global agriculture and nitrous oxide emissions. *Nature Climate Change, 2*(6), 410–416.
25. Scheer, C., Wassmann, R., Butterbach-Bahl, K., Lamers, J. P. A., & Martius, C. (2009). The relationship between N2O, NO, and N2 fluxes from fertilized and irrigated dryland soils of the Aral Sea Basin, Uzbekistan. *Plant and Soil, 314*(1–2), 273–283.
26. Green House Protocol. (2015). Global warming potential values (AR5). *Greenhouse Gas Protocol, 2014*(1995), 2–5.
27. Le Treut, H., et al. (2007). Historical overview of climate change. In: S. Solomon, S, & Qin, D., et al. (Eds.), *Climate change 2007*: the physical science basis. Cambridge: Cambridge University Press (pp. 93–127).
28. Bockman, O. C., & Olfs, H. W. (1998). Fertilizers, agronomy, and NO. *Nutrient Cycling in Agroecosystems, 52*(2/3), 165–170. Dordrecht.
29. Signor, D., et al. (2013). N_2O emissions due to nitrogen fertilizer applications in two regions of sugarcane cultivation in Brazil. *Environmental Research Letters, 8*(1), 1–9. Bristol.
30. Snyder, C. S., et al. (2009). Review of greenhouse gas emissions from crop production systems and fertilizer management effects. *Agriculture, Ecosystems & Environment, 133* (3–4), 247–266. Amsterdam.
31. Akiyama, H., Tsuruta, H., & Watanabe, T. (2000). N2O and NO emissions from soils after the application of different chemical fertilizers. *Chemosphere: Global Change Science, 2*(3–4), 313–320. Oxford.
32. Khalil, K., Mary, B., & Renault, P. (2004). Nitrous oxide production by nitrification and denitrification in soil aggregates as affected by O concentration. *Soil Biology & Biochemistry, 36*(4), 687–699. Oxford.

33. Davidson, E. A., & Swank, W. T. (1986). Environmental parameters regulating gaseous nitrogen losses from 2forested ecosystems via nitrification and denitrification. *Applied and Environmental Microbiology*, *52(*6), 1287–1292. Baltimore.
34. Liu, C., et al. (2011) Effects of irrigation, fertilization and crop straw management on nitrous oxide and nitric oxide emissions from a wheat-maize rotation field in northern China. *Agriculture, Ecosystems and Environment*, *140*(1–2), 226–233. Amsterdam.
35. Zhang, J., & Han, X. (2008). NO emission from the semiarid ecosystem under mineral fertilizer (urea and superphosphate) and increased precipitation in northern China. *Atmospheric Environment*, *42*(2), 291–302. Oxford.
36. Baggs, E. M., et al. (2000). Nitrous oxide emission from soils after incorporating crop residues. *Soil Use and Management*, *16*(2), 82–87. Oxford.
37. Brentrup, F., et al. (2000). Methods to estimate on-field nitrogen emissions from crop production as an input to LCA studies in the agricultural sector. *The International Journal of Life Cycle Assessment*, *5*(6), 349–357. Berlin.
38. Denmead, O. T., et al. (2010). Emissions of methane and nitrous oxide from Australian sugarcane soils. *Agricultural and Forest Meteorology*, *150*(6), 748–756. Amsterdam.
39. Ruser, R., et al. (2006). Emission of N2O, N2 and CO from soil fertilized with nitrate: effect of compaction, soil moisture and rewetting. *Soil Biology & Biochemistry*, *38(*2), 263–274. Oxford.
40. Stevens, R. J., & Laughlin, R. J. (1998). Measurement of nitrous oxide and di-nitrogen emissions from agricultural soils. *Nutrient Cycling in Agroecosystems*, *52(*2–3), 131–139. Dordrecht.
41. Thomson, A. J., et al. (2012). Biological sources and sinks of nitrous oxide and strategies to mitigate emissions. *Philosophical Transactions of The Royal Society Biological Sciences, London, 367*(1593), 1157–1168.
42. Bremner, J. M. (1997). Sources of nitrous oxide in soils. *Nutrient Cycling in Agroecosystems*, *49*(1–3), 7–16. Dordrecht.
43. Chapuis-Lardy, L., et al. (2007). Soils, a sink for N_2O?: a review. *Global Change Biology*, *13* (1), 1–17. Oxford.
44. Ciampitti, I. A., Ciarlo, E. A., & Conti, M. E. (2008). Nitrous oxide emissions from soil during soybean (*Glycinemax* (L.) Merrill) crop phenological stages and stubbles decomposition period. *Biology and Fertility of Soils*, *44*(4), 581–588. Berlin.
45. Cameron, K. C., et al. (2013). Nitrogen losses from the soil/plant system: a review. Annals of Applied Biology, *162*(2), 145–173. Warwick.
46. Davidson, E. A., et al. (2000). Testing a conceptual model of soil emissions of nitrous and nitric oxides. *Bioscience*, *50*(8), 667–680. Washington, DC.
47. Groffman, P. M., et al. (2009). New approaches to modelling denitrification. *Biogeochemistry*, *93*(1–2), 1–5. Dordrecht.
48. Cavigelli, M. A. & Robertson, G. P. (2001). Role of denitrifier diversity in rates of nitrous oxide consumption in a terrestrial ecosystem. *Soil Biology & Biochemistry*, *33*(3), 297–310. Oxford.
49. Baggs, E. M., Chebii, J., & Ndufa, J. K. (2006). A short-term investigation of trace gas emissions following tillage and no-tillage of agroforestry residues in western Kenya. *Soil & Tillage Research*, *90(*1–2), 6976. Amsterdam.
50. Baggs,E. M., et al. (2003). Nitrous oxide emissions following application of residues and fertilizer under zero and conventional tillage. *Plant and Soil*, *254(*2), 361–370. Dordrecht.
51. Huang, Y., et al. (2004). Nitrous oxide emissions as influenced by amendment of plant residues with different C: N ratios. *Soil Biology & Biochemistry*, *36*(6), 973–981. Oxford.
52. Six, J., et al. (2004). The potential to mitigate global warming with no-tillage management is only realized when practiced in the long term. *Global Change Biology*, *10(*2), 155–160. Oxford.
53. Van Kessen, C., et al. (2013). Climate, duration, and N placement determine NO emissions in reduced tillage systems: a meta-analysis. *Global Change Biology*, *19*(1), 33–44. Oxford.

54. Mosier, A. R. (2001). Exchange of gaseous nitrogen compounds between agricultural systems and the atmosphere. *Plant and Soil*, *228*(1), 17–27. Dordrecht.
55. Liu, X. J., et al. (2005). Tillage and nitrogen application effects on nitrous and nitric oxide emissions from irrigated corn fields. *Plant and Soil*, *276*(1–2), 235–249. Dordrecht.
56. Hellebrand, H. J., Scholz, V., & Kern, J. (2008). Fertilizer induced nitrous oxide emissions during energy crop cultivation on loamy sand soils. *Atmospheric Environment*, *42*(36), 8403–8411. Oxford.
57. Chen, S. T., Huang, Y., & Zou, J. W. (2008). Relationship between nitrous oxide emission and winter wheat production. *Biology and Fertility of Soils*, *44*(7), 985–989. Berlin.
58. Ma, B. L., et al. (2010). Nitrous oxide fluxes from corn fields: on-farm assessment of the amount and timing of nitrogen fertilizer. *Global Change Biology*, *16*(1), 156–170. Oxford.
59. Liu, X. J., et al. (2006). The impact of nitrogen placement and tillage on NO, N2O, CH4 and CO fluxes from a clay loam soil. *Plant and Soil*, *280*(1–2), 177–188. Dordrecht.
60. Shaviv, A. (2001). Advances in controlled-release fertilizers. *Advances in Agronomy*, *71*(1), 1–49. San Diego.

Chapter 2
Gas Adsorption and Urea Formation Using Magnetically Induced Environment

2.1 Introduction

Urea being an organic compound carries a chemical formula of $CO(NH_2)_2$ called carbamide. It contains the highest amount of nitrogen content among all nitrogenous granular-based fertilizers, making it the most produced and consumed and produced. However, its high water solubility properties create leaching issues which has attracted attention particularly from environmentalists. It is very suitable to be used as a fertilizer solution, which is commonly used as a foliar feed and could provide better efficiency if properly applied. Crop growth is hindered by the most common impurity of synthetic urea referred to as biuret [1].

Besides solubility in water, the production of urea requires high energy consumption due to its high-temperature and high-pressure requirements. To ensure a high yield production of urea, catalysts are needed. In short, synthetic urea can be produced by reacting carbon dioxide with synthetic ammonia in the presence of catalysts. It can be mostly found in solid granules or prills.

Fixed nitrogen from the air is the most important component in producing this nitrogenous-based fertilizer. It was found that in the process of developing nitrogen fixation, numerous principles of chemical and high-pressure processes have been simplified for efficiency reasons, hence, this has emerged in the chemical engineering field as well. Prior to the emergence of the synthetic nitrogen fixation, agricultural nitrogen fertilizers are primarily involved in a number of sources such as various waste and manure or decay products, and ammonium sulfate, considered as the secondary product extracted from coal used for cooking [2].

It is worthy to note that due to their energy-intensive industry, catalysts are required to reduce the activation energy. Catalysis increases the chemical reaction rate of a process. This saves energy as activation energy is reduced. It is an early procedure that must be performed before a reaction occurs. It includes the reduction of metallic oxide catalyst to reduced metallic phase caused by hydrogen as reducing

N. Yahya, *Green Urea*, Green Energy and Technology,
https://doi.org/10.1007/978-981-10-7578-0_2

gas. In the history of, ammonia synthesis, urea feedstock is considered among the most significant discoveries.

Ammonia is also a type of nitrogenous fertilizer. Many scientific journals focused on ammonia catalytic synthesis have been published even though it has been more than one century after the first demonstration of ammonia synthesis [2–21] through Haber–Bosch or other processes. Novel methods have been proposed such as the utilization of electromagnetic and magnetic energy a new technique with the presence of heterogeneous catalysis [2–6].

2.2 History of Ammonia and Urea Synthesis

Haber is a German scientist who found stupendously distributed osmium (Os) catalyst, producing 8% NH_3 at a pressure of 175 bars at around 600 °C in the year 1909 [7]. Haber discovered that uranium catalysts could be used to produce urea. He built a new laboratory plant in July, 1990, and had established the creation of 80 g of NH_3/hr with this new apparatus.

Nevertheless, based on the profit perspective, the catalysts Haber used were found to be unstable, and the reaction was found to be economically unfeasible. Haber along with Carl Bosch invented a cheaper and larger-scaled catalytic synthesis of ammonia from elemental hydrogen and nitrogen gases as reactants [7], hence the birth of a new technology for ammonia large-scaled synthesis.

A more systematic investigation considering the law of mass action kinetics and chemical equilibrium was performed by several chemists in the nineteenth century. From the equilibrium data, a small percentage of ammonia was obtained without any catalyst application apparently measured at normal pressure, and was kept well under 300 °C. Subsequently, the enhancement of the equilibrium conditions was performed by increasing the pressure and temperature.

However, this resulted in low ammonia yield despite elevated pressure of 75 bars and with an operating temperature up to 600 °C. Subsequently, Haber attempted to recycle the gas over the catalyst by this dynamic approach mixed with a fresh input of gas. Subsequently, the mixture was reprocessed under pressure and has become the groundwork for the technical process of ammonia production [3].

Indeed, a collaboration of Bosch and Haber has created the profitable intensive pressure NH_3 synthesis by the year 1913 [4]. The German chemical giant BASF (Badashe Analine und Soda Fabric) situated in Oppau, was the earliest profitable plant with a capacity of 30 tons/day [8]. Recently, some considerable modifications have been addressed to this energy-intensive process so that it yields a price that does not increase as rapidly as energy prices.

A Dutch scientist, Herman Boerhaave, first discovered urea in urine in 1727 [9]. French chemist by the name of Hillaire Rouelle later discovered urea in 1773 [10]. Frederick Wohler, an organic chemist had synthetically formulated urea. It was formed by dehydrating ammonium carbamate from a reaction of carbon dioxide and

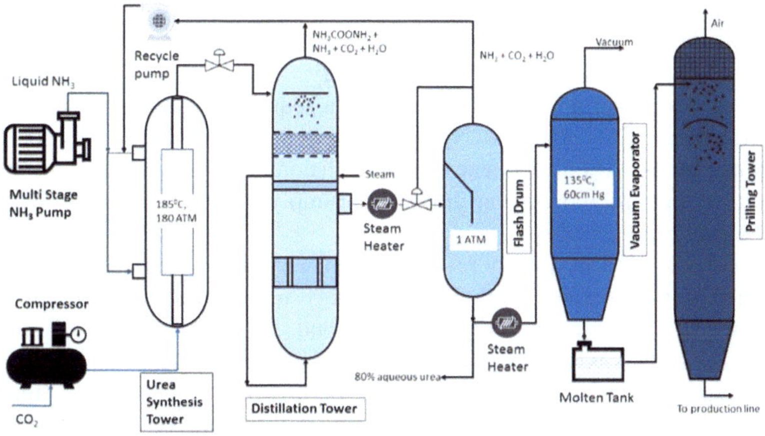

Fig. 2.1 Schematic diagram of production of urea from NH_3 and CO_2 [11]

ammonia under high pressure and high-temperature conditions which were first implemented in 1870 [10].

Initially, the formation of ammonium carbamate takes from the reaction of NH_3 and CO_2 under high temperature and pressure, followed by urea conversion. The ammonium carbamate then decomposes into urea and water (H_2O). Equations (2.1) and (2.2) show the chemical equation of the urea synthesis process [12]. The first reaction almost goes to completion at 135–200°C and 180–200 bars [13, 14]. Since the decomposition reaction is a slow process, the rate of urea production is determined by the decomposition reaction. The operating temperature of the decomposition reaction is 135°C and 35 bars of pressure [13, 14]. The endothermic reaction of urea formation maintains ammonia amounts in excess so as to shift the equilibrium toward the formation of urea. Urea processes produce an aqueous solution containing 70–87% urea [15]. It is easy to produce urea as prills or granules [16]. Figure 2.1 shows a schematic diagram of the conventional urea synthesis process.

$$2NH_3 + CO_2 \rightleftharpoons H_2N - COONH_4 \quad (2.1)$$

$$H_2N - COONH_4 \rightarrow (NH_2)_2CO + H_2O \quad (2.2)$$

2.3 Magneto-Dynamics and Catalytic Activity

Chemical reactions have a remarkably direct relationship with magnetic fields [17, 18]. The influence of an externally applied magnetic field to a chemical reaction has been investigated in the related literature. The singlet to triplet conversion state of carbon dioxide and other gases is the least energy required to start a chemical reaction is the energy of activation (E_a) [19]. It is well-known that the presence of catalysts is able to reduce the activation energy.

The mechanism of a chemical reaction is a sequence of actual events in which reactant species: CO_2, H_2, and N_2 are converted into urea. The reaction rate is determined by the overall frequency of the collisions among the three gases namely, nitrogen, hydrogen, and carbon dioxide or simply ammonia and carbon dioxide with the iron oxide catalysts.

A higher collision frequency results in a greater reaction rate. Urea product will be formed if there exists ahigh kinetic energy of the reactant gases that cause weakening subsequently the breaking of a bond. In order to effectively initiate a reaction, the collisions must have sufficient energy to break the bond. The minimum energy must be obtained initially. On top of this, the reacting gases must also collide with proper orientation [20]. In any collision involving unsymmetrical species, the manner in which the molecules hit each other is important whether or not a reaction will occur.

The basic knowledge of science in thermodynamics is a statistical analysis of vibration and the corresponding drive in populations of atoms or molecules. The materials with unpaired electrons will generally display magnetic properties. According to the rules of quantum electrodynamics, when a molecule is placed under a magnetic field, the unaligned spins of electrons will be aligned parallel or antiparallel in the direction of the applied magnetic field [21]. However, in the non-existence of a magnetic field, this similar number may have closely unrestricted orientation of movement. The magnetic field effects on the thermodynamics properties of a chemical reaction have been systematically studied [22].

By applying the magnetic fields to the ferromagnetic metal-hydrogen system $LaCo_5$-H_2, the dissociation of H_2 from the metal hydride occurs, associated with an enhancement of hydrogen pressure equilibrium. It has been proven that the equilibrium constant of a chemical reaction is influenced by magnetic fields. Apart from this, the equilibrium constant is also influenced by free energy, heat, and entropy of the reaction. This part illustrates the change in chemical equilibrium with the presence of a magnetic field. The ultimate aim is to consider the possibility of ammonia and urea formation using magnetically induced conditions.

Consider a chemical system where k represents the types of ideal gases that mutually react with a stoichiometric balance and coefficients v_i (i = 1 ~ k) under the effect of a magnetic field. Chemical equilibrium conditions can be written as per Eq. (2.3) [22]:

$$\sum_{i}^{k} v_i \mu_i = 0 \tag{2.3}$$

The$\mu_i^{(0)}$and $\mu_i^{(m)}$ terms of the ith gas contain the non-magnetic and magnetic terms as per Eq. (2.4):

$$\mu_i = \mu_i^{(0)} + \mu_i^{(m)} = \mu_i^* + RT\ln\frac{P^i}{P^*} - \int_0^H \mu_0 \boldsymbol{m}_i d\boldsymbol{H} \tag{2.4}$$

where μ_i is the normal chemical potential at temperature T and the normal pressure P^* in the absence of magnetic fields. P_i and $\boldsymbol{m}_i$ are the fractional pressure and the molar magnetization of the ith component. μ_0 is the permeability of free space. The integration was performed with respect to the applied magnetic field strength $\boldsymbol{H}$ at a continuous temperature T.

The equilibrium constant (K_p) is resultant from Eqs. (2.5) and (2.6) as per Eq. (2.5):

$$\ln K_P = \sum_{i=1}^{k} v_i \ln\frac{P^i}{P^*} = -\frac{1}{RT}\sum_{i=1}^{k} v_i\left(\mu_i^* + \mu_i^{(m)}\right) \tag{2.5}$$

Equation (2.5) leads to the magnetic field induced causing change in the equilibrium constant (2.6)

$$\ln K_P^{[H]} - \ln K_P^{[0]} = -\frac{1}{RT}\left(-\sum_{i=1}^{k} v_i \int_0^H \mu_0 \boldsymbol{m}_i d\boldsymbol{H}\right) = -\frac{1}{RT} g^{(m)} \tag{2.6}$$

The superscripts [H] and [0] symbolize the K_p in the applied magnetic field and in zero fields, respectively. The other thermodynamic quantities such as $g^{(m)}$ are the change of magnetic free energy per unit reaction.

The magnetic free energy change per unit reaction $g^{(m)}$ is given by Eq. (2.7):

$$g^{(m)} = -\sum_{i=1}^{k} v_i \int_0^H \mu_0 \boldsymbol{m}_i d\boldsymbol{H} = -\sum_{i=1}^{k} v_i (\mu_0 \boldsymbol{m}_i \boldsymbol{H}), \tag{2.7}$$

and the magnetic field strength is written as Eq. (2.8):

$$\boldsymbol{H} = \frac{B}{\mu_0} - M \tag{2.8}$$

where $\boldsymbol{H}$ is the applied magnetic field strength, B is the magnetic induction strength in the unit of tesla, and M is the magnetization field (also denoted as $\boldsymbol{m}_i$). Substituting Eq. (2.7) leads to Eqs. (2.9) and (2.10):

$$g^{(m)} = -\sum_{i=1}^{k} v_i \mu_0 \left[\boldsymbol{m}_i \left(\frac{B}{\mu_0} - \boldsymbol{m}_i\right)\right] \tag{2.9}$$

$$g^{(m)} = -\sum_{i=1}^{k} v_i [\boldsymbol{m}_i B - \boldsymbol{m}_i \mu_0] \tag{2.10}$$

In addition, the terms ΔH^0 ΔG^0 ΔS^0 T and $\boldsymbol{H}$ denote heat of reaction, the free energy of reaction, the entropy of reaction, the function of temperature and magnetic field strength, respectively. These physical quantities follow the standard theory of chemical thermodynamics that deals with the magnetic field changes induced in their respective thermodynamic quantities, as per Eqs. (2.11, 2.12, and 2.13) [22]:

$$\Delta H^{[H]} - \Delta H^{[0]} = g^{(m)} - T\frac{d}{dT}g^{(m)}, \tag{2.11}$$

$$\Delta G^{[H]} - \Delta G^{[0]} = g^{(m)}, \tag{2.12}$$

$$\Delta S^{[H]} - \Delta S^{[0]} = -\frac{d}{dT}g^{(m)}, \tag{2.13}$$

Equations (2.11), (2.12), and (2.13) are the general formulae for the magneto-thermodynamic effects in chemical reactions.

Equation (2.9) shows that the magnetic induction strength B is proportional to the change of per unit reaction $g^{(m)}$ of magnetic free energy. Therefore, increasing B tends to increase $g^{(m)}$. Moreover, Eq. (2.10) also denotes that the Gibbs free energy of a system can be increased by applying an external magnetic field to the system and that $\Delta G^{[H]}$ is proportional to $g^{(m)}$.

This means that the ammonia and urea synthesis yield can be improved by increasing the applied magnetic induction strength via the magnetic induction method. The magnetically induced chemical reaction will be elaborated in the subsequent chapter, where by a sequence of tests was performed on reactors for ammonia and urea synthesis.

2.3.1 Heterogeneous Catalyst

Catalysis is the increase of the rate of a chemical reaction with the presence of catalysts. The catalysts are not consumed in this reaction and only a small amount of it is needed in any reaction [23]. The reaction can be classified as either heterogeneous or homogeneous, which largely depends on the phase of the reactant molecules. Ammonia and urea synthesis adopts a heterogenous catalytic reaction which is paramount and important in the chemical industry. The initial step in the heterogeneous catalysis of ammonia synthesis is the adsorption of reactants, where the hydrogen and nitrogen gases bind to the catalyst surface.

Adsorption occurs due to the atoms or ions at the surface of the catalysts in this case, iron being extremely reactive. A high electron density at the tip of an iron wire has an unused bonding capacity that can be used to bond molecules from the reactant gases, nitrogen and hydrogen to the catalyst surface [23]. There are two parts in the adsorption process of ammonia synthesis, i.e., physisorption (physical adsorption) and chemisorption (chemical adsorption).

Physisorption involves intermolecular forces (van der Waals forces), where the electronic orbital patterns of the species involved do not undergo any significant changes [24]. The enthalpy change associated with physisorption is typically 40 kJ/mol or less [25]. In this stage, the reactant gases flow into a reaction medium. The reactant gases then attach themselves to the catalyst surface due to an attraction force, i.e., the van der Waals force. Since the enthalpy change for physisorption is too small to cause the molecules bonds to cleave, the physisorbed molecules remain intact and can be easily freed. Eventually, multiple layers of adsorbed molecules are built up, resulting in zero activation energy [26].

Chemisorption involves the sharing of electrons between the adsorbate (gases) and adsorbent (catalyst). The forces involved are valance forces of the same kind as the formation of chemical compounds [24]. In the chemisorption process, the enthalpy change is typically over 80 kJ/mol and sometimes even greater than 400 kJ/mol [25]. In fact, during the chemisorption process, the adsorbing reactant gases split into atoms via the sharing or exchange of electrons among the reactants and catalyst surface.

Unlike physisorption, chemisorption occurs only when the adsorbate is in contact with the surface, as in a single-layer process [27]. A deposit of molecules might be adsorbed physically on top of an underlying chemisorbed layer, Physical adsorption may also occur on non-active sites of a substrate, while chemisorption occurs on the active sites [27] only. Furthermore, when nitrogen molecules are diffused on an iron catalyst surface, the triple bonds of the nitrogen molecules ($N{\equiv}N$) will donate their electrons to the incompletely engaged *d*-orbital of the iron (sigma (σ) donation) [28].

As a consequence, a strong chemical bond is formed in this σ donation, thus resulting in a smaller distance between the absorbed nitrogen molecules and the iron surface [28]. The iron possesses five parallel electrons in its *d*-orbital and requires five additional electrons to stabilize. Therefore, the excess electrons will be provided back to the vacant pi (π^*) orbital of the N_2 molecules (π-back donation). As a consequence, the nitrogen triple bonds are lengthened, weakened, and cleaved, resulting in the dissociation of N_2 molecules on the iron surface to become nitrogen atoms.

Figure 2.2 presents the mechanism of nitrogen chemisorption on the iron oxide catalyst surface. The hydrogen chemisorption occurs in the same fundamental manner. During the hydrogen chemisorption process, a lone 1 *s* electron of each H-atom coordinates with the *d*-suborbitals of the iron, forming a pair of chemisorption bonds. Although these new bonds are found to be more stable compared to the single covalent bond replaced, the subsequent H-atoms are capable of wandering along the surface due to the continuous range of the *d*-orbital conduction band of the catalyst [26]. Until this stage, all the chemisorbed hydrogen has reacted with the chemisorbed nitrogen on the adsorbent surface to form ammonia. Figure 2.3 illustrates the phenomenon of H_2 chemisorption over the iron-based catalyst surface.

After the hydrogen atom reacts with the nitrogen atom on the catalyst surface and forms ammonia, the intermediate strength of the surface bonds causes the desorption process of ammonia to occur. Consequently, this leaves active sites free

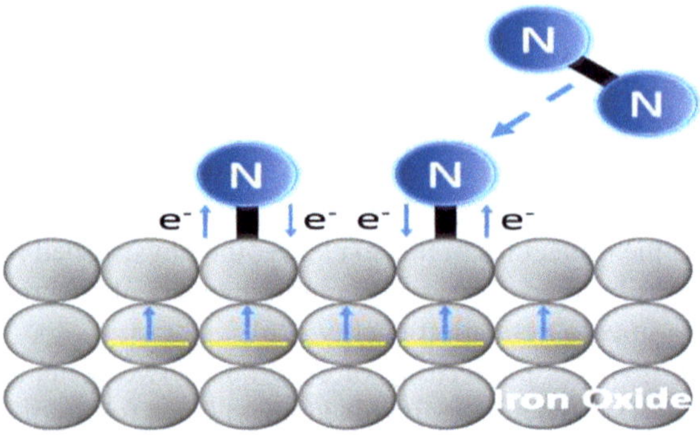

Fig. 2.2 Mechanism of nitrogen chemisorption on the iron oxide catalyst surface

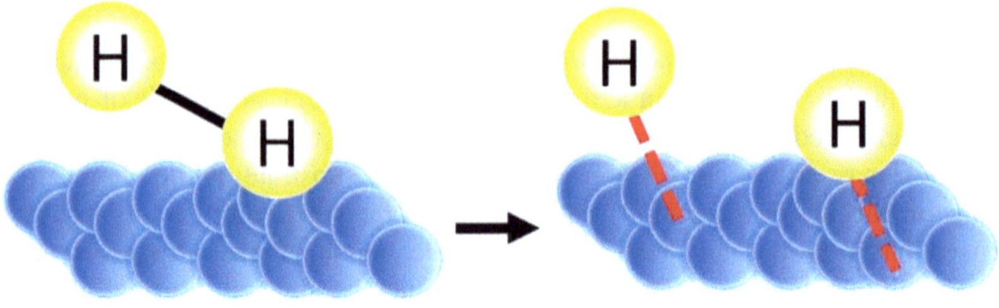

Fig. 2.3 Process of hydrogen chemisorption on iron-based catalyst surface [26]

for the incoming reactant molecules to occupy. Figure 2.4 summarizes the catalytic reaction in this section, adsorption and desorption, of the ammonia synthesis process. The covalent bonds of the molecules become crippled when the hydrogen and nitrogen molecules bind to the surface of the catalyst; eventually causing the molecules to dissociate. The hydrogen and nitrogen atoms, being highly reactive, bond together to form ammonia molecules and then leave the catalyst surface, indicating a desorption process (Fig. 2.4).

2.3.2 Density Functional Theory (DFT)

First principle density functional theory (DFT) is a computational quantum mechanics modeling methods applied to quantify the surface reaction mechanism. It is an extremely effective and extensively applied method for approximating the solution of Schrödinger equation for a system of electrons in materials. It can be followed to assess the ground level of several large electron systems, such as atoms, molecules, and clusters. In this work, plane-wave pseudo-potential calculations

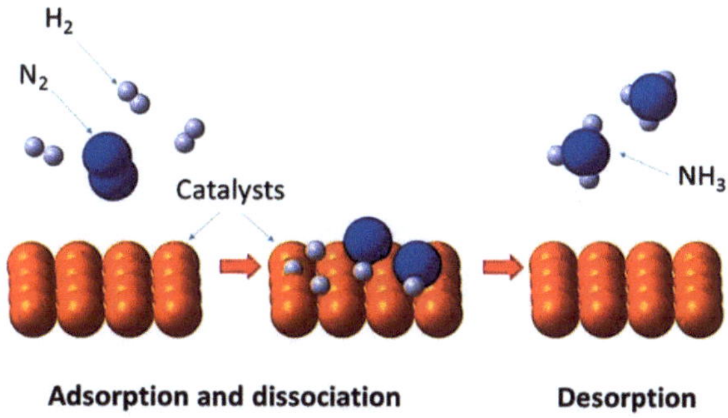

Fig. 2.4 Catalytic reaction in the ammonia formation [27]

have been performed using the Cambridge Serial Total Energy Package (CASTEP). This involves a study of surface chemistry structural features, band structure, density of states, and optical properties [29].

The sorbates involved in this study are H_2, N_2, and CO_2 gas molecules, whereas sorbents are CuO and Fe_2O_3 cleaved plane at 5 Å each. For the computational study, the desired materials in the form of their unit cells were used in the Material Studio software library, and were then converted into the desired cleaved surfaces. As CASTEP properties involve periodicity of structure, the modified nanocluster structures of CuO and hematite (Fe_2O_3) were then transformed into a vacuum slab [30].

The structure of CuO and Fe_2O_3 is geometrically optimized by CASTEP and then cleaved to an indexed plane of CuO and Fe_2O_3. The H_2, N_2, and CO_2 molecules were optimized by forcite module and adsorbed on CuO (111) and α-Fe_2O_3 (111) plane by an absorption locator. Energy calculations were performed to compute the electronic properties of the system.

The nanocatalysts cleaved surface electronic band structures, total and partial density of states (TPDOS), difference of total and partial spin density of states (TPSDOS) of alpha (up spin) and beta (down spin) electrons with the adsorption of gases were calculated. DOS-alpha (up spin) and DOS-beta (down spin) [31] calculations were executed for a spin-polarized system.

The first part of the simulation was done to evaluate the adsorption of carbon dioxide as a molecule on the surface of a catalyst. Its versatility in the choice of substrate (metallic) and adsorbate (atomic and molecular) makes it particularly valuable when information is not easily accessible with conventional experimental techniques. When the carbon dioxide evaluation was complete, the three gases adsorption was calculated. It is worthy to noted that the adsorption of the greenhouse gas on the surface of a catalyst may be evaluated as green. This is due to the fact that the understanding of this mechanism enables us to calculate and evaluate other chemical reactions as well. If carbon dioxide is adsorbed on the above catalysts, the possibility of the catalytic activity taking place is as high as explicitly explained in the above section. This section presents a thorough discussion on the change of its electronics properties with the adsorption of the gases on the catalyst. The change in net spin, band gap, and density of states was calculated and

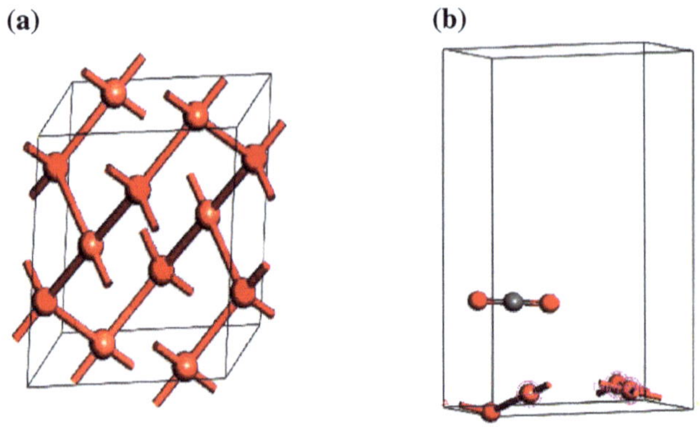

Fig. 2.5 Unit cell of **a** CuO (111) and **b** CuO (111) with CO_2 adsorption

evaluated. The results were subsequently validated with the experiments as validation of the preliminary concepts of magneto-dynamics.

The adsorption energy represents the difference in total energy of material before and after the adsorption of gases [32], and it is a consequence of a surface energy. It can be defined by Eq. 2.14.

$$\Delta E_a = E_{surface + gas\, molecule} - E_{surface} - E_{gas\, molecule} \tag{2.14}$$

From an energy point of view, the most likely favored sites have the largest adsorption energy, and are likely the most active sites. The largest negative values of E_{ads} for reactant gases propose that these molecules are significantly and spontaneously adsorbed onto α-Fe_2O_3 surfaces [33] Fig. 2.5a shows the (111) surface of CuO, and 5 (b) is the top of CuO with the adsorbed carbon dioxide gas molecules.

This section, discusses the change in density of electrons by analyzing the change in energy of electrons occupied by partial orbitals. Figure 2.6 represents the calculated band structure of (a) CuO (111) before adsorption of CO_2 and (b) CuO (111) after adsorption of CO_2 gas. The blue-dotted line represents the Fermi level. The adsorption of CO_2 on the CuO resulted in a reduction of the band gap from 1.48 to 1.13 eV, a reduction of 35% of the energy. This designates that by the gas adsorption, less energy is obligatory to move an electron from the valence level to the conduction band.

The reduction is approximately 35% of adsorption energy, which can be translated to a reduction of energy savings of 35%, considering carbon dioxide is a greenhouse gas.

The reduction in the band gap can also be well-understood on the basis of electronic structure of solids. When an adsorbate approaches a catalyst surface, the electronic states of the catalyst interact with an adsorbate energy level. The energy level expands and moves upon coupling with the wide metal s-band. Consequent coupling of the adsorbate level to the more localized d-states solidifies a splitting into a symmetric and an antisymmetric part of the adsorbate levels.

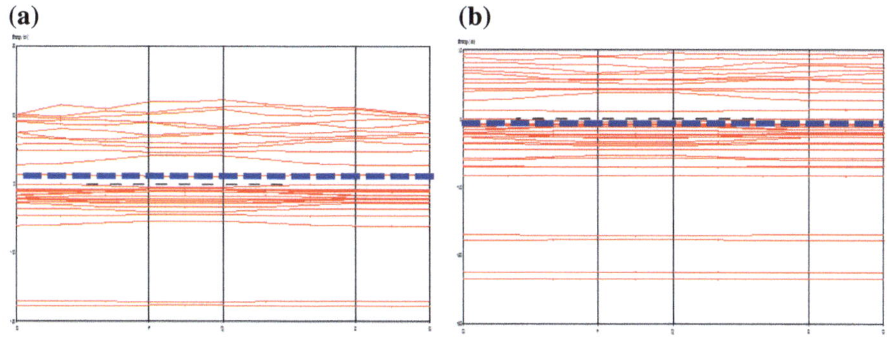

Fig. 2.6 Band structure of **a** CuO (111) before and **b** CuO (111) after CO_2 adsorption

In fact, number of up spins is equal to the number of down spin and just below Fermi level bands are presented in Fig. 2.7. The cleaved surface is broken and the possible presence of dangling bonds makes the surface reactive. The wave function of molecules of adsorbed carbon dioxide gas interacts with the wave functions of surface atoms through an exchange-correlation function, which results in the splitting of orbitals energy. The zero net spin indicates CuO as a non-magnetic element. Referring to Table 2.1, even though the net spin is zero for both CuO before and after adsorption of CO_2, the number of spins differ, 17 and 25, respectively, which constitute to about 32% of number of spins.

Figure 2.7 represents the electronic density of states of electrons (a) before and (b) after adsorption of gas molecules. (a) The maximum energy DOS of electrons at the Fermi level is not high, and the size is not as much as it is in Fig. 2.7b. The splitting in symmetric and antisymmetric orbitals of the surface can be well-understood by the electronic nature of the surface orbitals, and the filling of partially unfilled orbitals for a surface can be understood by the Fermi level.

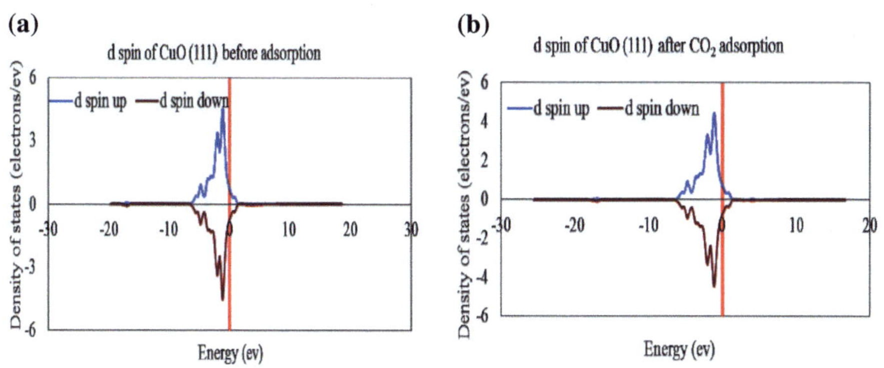

Fig. 2.7 d spin density of state (eV) of **a** CuO (111) before and **b** CuO (111) after CO_2 adsorption

Table 2.1 Bad gap properties post-absorption of CO_2 of different materials

Material	Band gap (eV)	Up spin	Down spin	Net spin	% decrease in band gap after absorption of CO_2
CuO	1.69	34	34	0	23.6
CuO (111)	1.48	17	17	0	
CuO (111) with absorbed CO_2	1.13	25	25	0	
ZnO	0.73	18	18	0	24.1
ZnO (101)	0.29	72	72	0	
ZnO(101) with absorbed CO_2	0.22	80	80	0	
α-Fe_2O_3	1.32	102	102	0	84
α-Fe_2O_3 (111)	0.12	34	34	0	
α-Fe_2O_3 (111) with absorbed CO_2	0.019	40	20	20	

When any reactant on the surface of hematite is adsorbed, electronic arrangements differ with the change in positions of atoms resulting in the intersecting of 1s1, $2p^3$, and $2p^2$ subatomic orbital of H_2, N_2, and CO_2 molecules with d-orbital of CuO (111). The intersection of electronic wave functions enhances the width of the energy density of states at the FL [34].

Phonon density of state is crystal vibration of atoms in a crystal lattice. Partial density of state (PDOS) of CuO (111) calculation was performed and presented in Fig. 2.8. Figure 2.8a and b present the PDOS of CuO (111) pre-and post-adsorption of CO_2 gas molecules.

After the adsorption of CO_2 gas on the surface of CuO (111), the change in s-, p-, and d-orbitals can be observed. When the CO_2 molecule and CuO (111) surface interacts, the respective states mixed and new states are created which have energy levels usually broadened and shifted with respect to the energy levels.

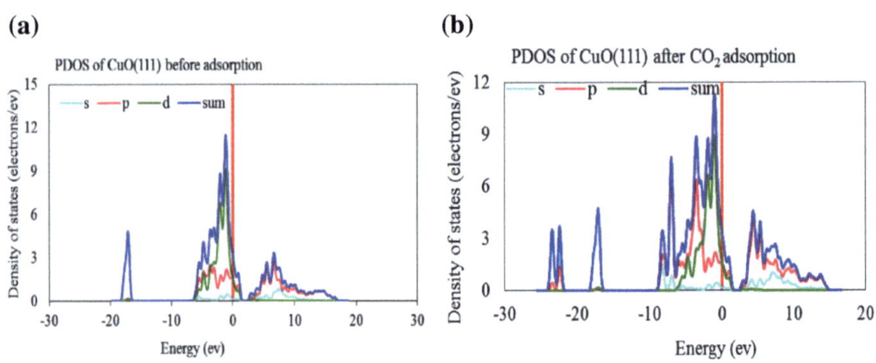

Fig. 2.8 Partial density of states (eV) of **a** CuO (111) before **b** CuO (111) after CO_2 adsorption

Observing the sum of partial DOS post-adsorption, it is obvious that there exists a higher number of states per interval of energy at different energy level. There are more number of states above 0 (eV) energy level after the adsorption of CO_2. Electron energy is also increased above 0 eV post-adsorption.

Gases interaction with the surface of magnetic nanocatalyst for the synthesis of ammonia and urea must be well-understood. The understanding of the adsorbed gases (in this case carbon dioxide) that might influence the magnetic properties and catalytic activity of the nanocatalyst is essential to move forward in the evaluation of the energy reductions hence saving that constitute to the green effect. Most chemical processes utilize catalysts to improve the production yield; this part of the work allows for a prediction of how gas such as carbon dioxide would change the properties (particularly the magnetic properties) of the catalysts.

The system was geometry optimized using the Broyden–Fletcher–Goldfarb–Shanno (BFGS) procedure to attain lattice parameters and atomic locations corresponding to the lowest energy configuration. A rhombohedral structure is used for ground state calculations. Figure 2.9a presents the geometry of Fe_2O_3 (111) and (b) gives the surface geometry of Fe_2O_3 (111) with the CO_2.

Figure 2.10 describes the electronic band structure that calculated the range of energies of electron for Fe_2O_3 (111) without and with the adsorption of CO_2 gas. It should be noted that the reduction of energy with the adsorption of CO_2 is obvious, that is, from 0.12 eV to 0.019 eV which constitutes to 84% (Table 2.1). The reduction in band gap can also be well-understood on the basis of the electronic structure of solids.

When an adsorbate approaches the catalysts surface which is the absorbent, the electronic states of the substrate interact with an adsorbate energy level the density of states of Fe_2O_3 (111) is presented in Fig. 2.11. The number of spin up, 40, (in d-orbital is not equal to the number of spin down, 20; hence, Fe_2O_3 (111) post-adsorption of Fe_2O_3 (111) is in magnetic state (Table 2.1).

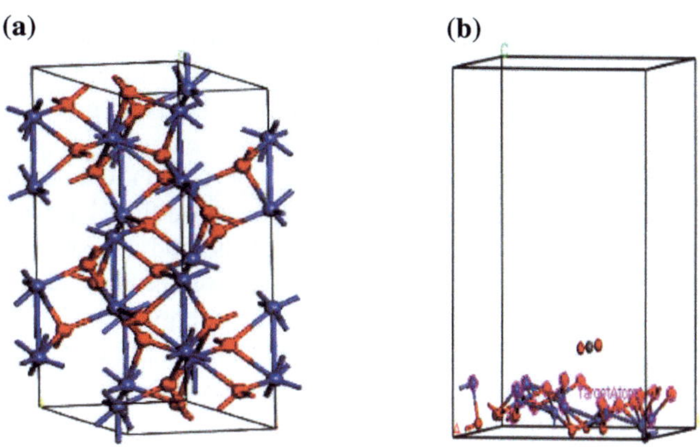

Fig. 2.9 Unit cell of **a** Fe_2O_3 (111) and **b** surface Fe_2O_3 (111) with CO_2

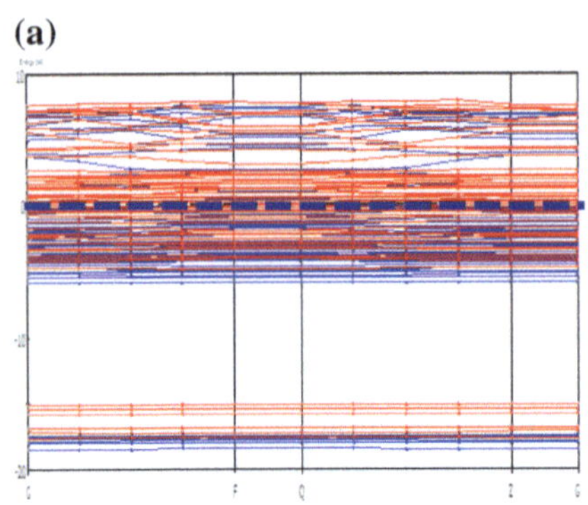

Fig. 2.10 Band structure (energy (eV) of **a** Fe_2O_3 (111) **b** Fe_2O_3 (111) with adsorbed CO_2

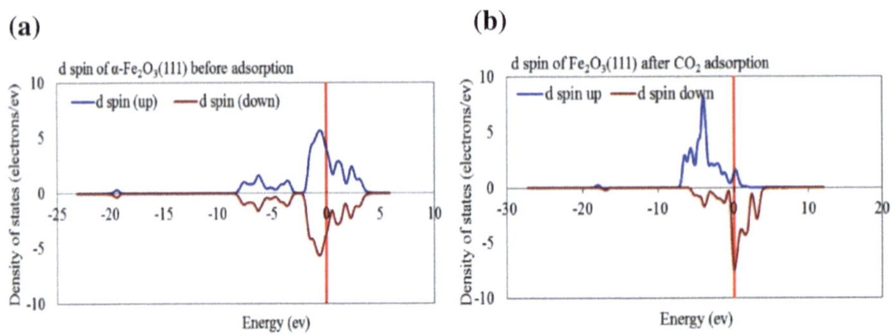

Fig. 2.11 d spin density of states (eV) **a** number of up spin in d-orbital is not equal to the down spin hence Fe_2O_3 is in magnetic state after adsorption of **b** Fe_2O_3 (111) with adsorbed CO_2

It is noteworthy that the up and the down spins are equal for both Fe_2O_3 (111) and CuO (111), 34 and 17, respectively. CuO is conductive, while Fe_2O_3 is magnetic. The FL is shifted to the valance band, and the band gap is compact after the adsorption of the greenhouse gas.

Figure 2.12 presents the density of states of Fe_2O_3 (111) without and with CO_2 adsorption. The 3d and 4s levels split the most because they are farther from the nucleus so they interact first. Lower levels (2p, 2s, and 1s) show less splitting because they are closer to the nucleus. More electrons mean more splitting in the orbitals, and these levels are so close that they form energy bands [34]. In this case, when CO_2 molecules interact with the Fe_2O_3 (111) surface, the 3d and 4s levels of α-Fe_2O_3 (111) couple with the 1s and 2p orbitals of CO_2 molecules which results in the splitting of energy levels.

In Fig. 2.12a, the splitting of energy levels is maximum in the range of energy of −27–5 eV. DOS at 2s level is very high, 13 eV. As previously mentioned the

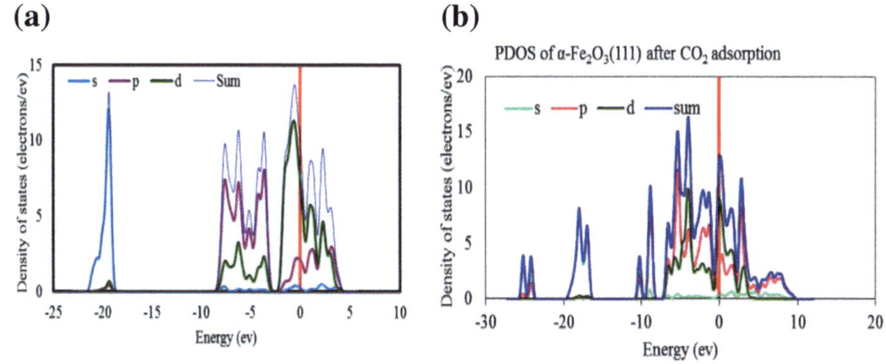

Fig. 2.12 PDOS of **a** Fe_2O_3 (111) **b** Fe_2O_3 (111) with CO_2 adsorbed

cleaved surfaces have broken and dangling bonds making the surface reactive. The wave function of molecules of adsorbed gas interacts with the wave functions of surface atoms through exchange-correlation function, resulting in the splitting of orbitals energy. When molecules and a surface interact, the respective states mix and new states are created which have energy levels usually broadened and shifted with respect to the energy levels found in the uncoupled systems [35].

As seen in Table 2.1, the band gap decreases as the material absorbs CO_2.

2.4 Simulation and Experimental Evaluation of NiO Nanoparticle for Urea Synthesis

In reality, the adsorption of three gases forming urea, namely, H_2, N_2, and CO_2 need to be understood. The nature of magnetic systems is quantum mechanical due to its quantized energy levels of electrons. The exchange interaction among electrons and magneto crystalline anisotropy emerges from relativistic interactions of subatomic electronic orbitals with the lattice, spin–orbit coupling, a pure effect of quantum mechanics [36].

The Heisenberg model of spin captures the critical physics of a magnetic material at the subatomic level, where the energetics of a structure of cooperating atomic moments are given by a Hamiltonian of spin (which disregards non-magnetic belongings such as the Coulomb term). Studies on evaluation of Fe_2O_3 as a catalyst were able to induce magnetic moment to 3.33 μB indicating the gas adsorption was able to improve the magnetic properties of the catalysts [37]. Carbon dioxide has a significant role in urea synthesis, and may potentially be able to reduce the global warming potentials.

The adsorption energy and spin modifications of adsorbed gases of favorable morphology and composition of nanocatalysts play a vital role in chemical activity, rate, and yield for specific fertilizers synthesis. The sorbates involved in this study are H_2, N_2, and CO_2 gas molecules, and sorbents are considered to be NiO (111) nanocatalyst cleaved surface. For the first principle DFT study evaluation a metal oxide-NiO, cubic unit cell structure was imported from the material studio library and then transformed into a NiO (111) cleaved surface with a super cell symmetry of A = 1, B = 1, followed by a 3D vacuum slab. This is because the CASTEP properties involve periodicity of structure [16]. Subsequently, by using an Adsorption Locator module, the H_2, N_2, and CO_2 molecules were constructed individually, cleaned, symbolically labeled, and then forcite geometry optimized. Then NiO (111) nanocatalyst cleaved surface was selected as the target surface, and forcite geometry optimized sorbate molecules (H_2, N_2, and CO_2) were loaded at a set distance of 7 Å. Finally, the setup was excuted for adsorption energy calculations of all sorbate molecules. The electronic band structures, total and partial density of states (TPDOS), difference of total and partial spin density of states (TPSDOS) of alpha (up spin) and beta (down spin) electrons, and number of bands were quantified for isolated NiO unit cell; isolated NiO (111) cleaved surface, NiO (111) with CO_2 and NiO (111) with CO_2, H_2, and N_2 molecules within identical vacuum slab at 7 Å, respectively. For band structures and DOS calculations, CASTEP module was chosen from the menu bar to display the CASTEP calculations dialog. Then following the setup tab band structure and DOS tasks were selected from the drop-down list. The quality of calculation was fixed to medium for precise calculations, and the exchange-correlation function was adjusted to GGA (generalized gradient approximation) and PBE (*Perdew Burke Ernzerh*). We checked the spin-polarized option, use formal spin as initial checkboxes and optimizing total spin checkbox, respectively. As calculations were executed for DOS for a spin-polarized system with a nonzero/formal initial spin; the two charts automatically appeared as DOS-alpha (up spin) and DOS-beta (down spin) [6].

2.4.1 DFT Simulation Adsorption Study of Nickel Oxide NiO (111) with H_2, N_2, and CO_2 Molecules

The adsorption energy values of urea reactant molecules (CO_2, H_2, N_2) were calculated by using Accelry adsorption anneal module. The desired molecular structures were constructed and then geometrically optimized by using forcite module. These molecules were then loaded to a built-in vacuum slab over NiO (111) nanocatalyst cleaved surface at distance of 7 Å as shown in Fig. 2.13d. The calculated average adsorption energy values of CO_2, N_2, and H_2 molecules were found to be –3.202 (kcal/mol), −1.758 (kcal/mol) and −2.644 (kcal/mol), respectively. The more negative value of CO_2 molecule reflects the comparatively appropriate physisorption as compared to H_2 and N_2 molecules.

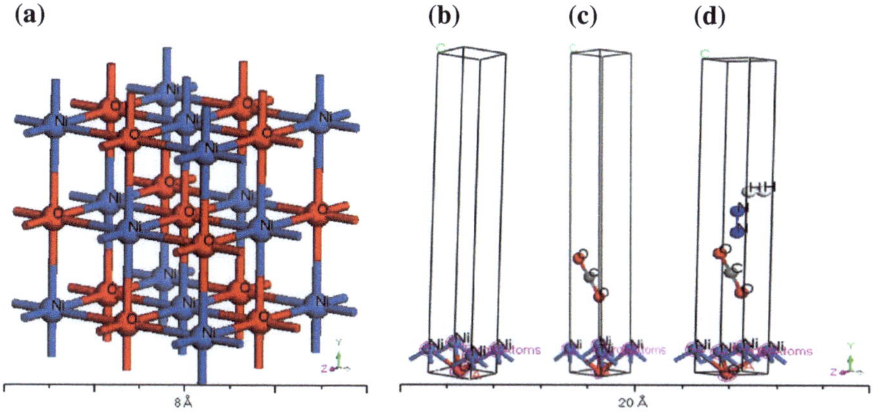

Fig. 2.13 **a** NiO unit cell **b** NiO (111) (**c** and **d**) NiO (111) in 15 Å vacuum slab thickness with CO_2 and with H_2, N_2, and CO_2 in 15 Å vacuum slab thickness, respectively

2.4.2 DFT Simulation Study of Electronic Band Structure and Partial Density of States of NiO Unit Cell, NiO (111, NiO (111), CO_2 and NiO (111), CO_2, H_2, N_2 Molecules

2.4.2.1 DFT Simulation Study of Electronic Band Structure and Partial Density of States of NiO Unit Cell

This section, presents a comparative study of electronic band structure, partial density of states, and partial net spin density of states of all subatomic orbitals of NiO (Ni = 3d8, 4s2, O = 2s2, 2p4) and reactant molecules (C = 2s2, 2p2, N = 2s2, 2p3, H = 1s1) across the fermi level (FL) for NiO unit cell, NiO (111), and NiO (111) with CO_2, NiO (111) with CO_2, H_2, and N_2 molecule at 7 Å. Figure 2.13a–d shows the simulation modeling of NiO unit cell, NiO (111), NiO (111) with CO_2 and NiO (111) with CO_2, H_2, and N_2, respectively.

The electronic band structure is shown in Fig. 2.14a–c, respectively. The blue–red lines in the electronic band structure of NiO unit cell represent the electronic states of alpha (up) and beta (down) electrons for spin-polarized calculations across the Fermi level (FL). Most of the electronic states of alpha–beta electron are above the Fermi level within an energy range of 0–40 eV, and significantly occupied electronic states are observed across the FL between 5 and −10 eV. The band structure for conduction and valance bands across the FL is symmetrical across the hove singularity/symmetry points/Brillion zone (W, L, G, X, W, and K).

Figure 2.15a shows a corresponding partial orbital density of states of subatomic orbitals (s, p, d, and total) of NiO (Ni = 3d8, 4s2 and O = 2s2, 2p4) unit cells. The maximum PDOS lie between energy ranges of −10–20 eV across the FL with

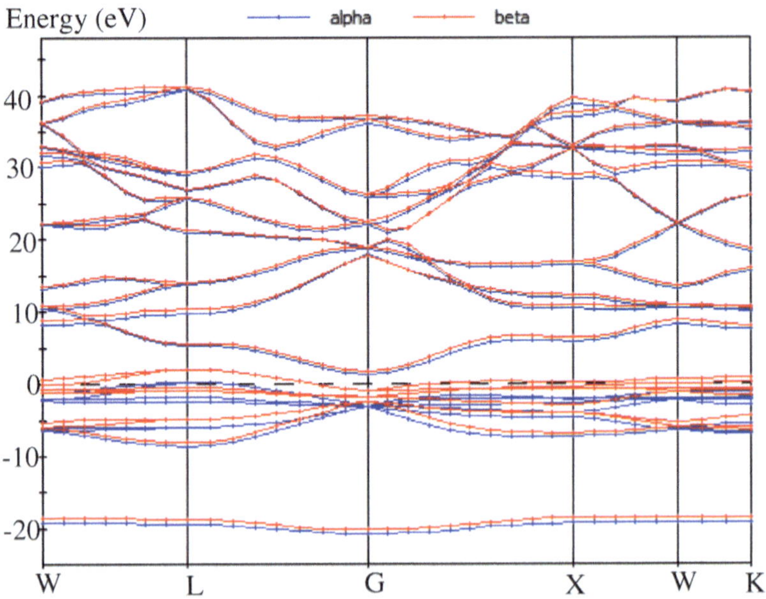

Fig. 2.14 Electronic band structure of NiO unit cell

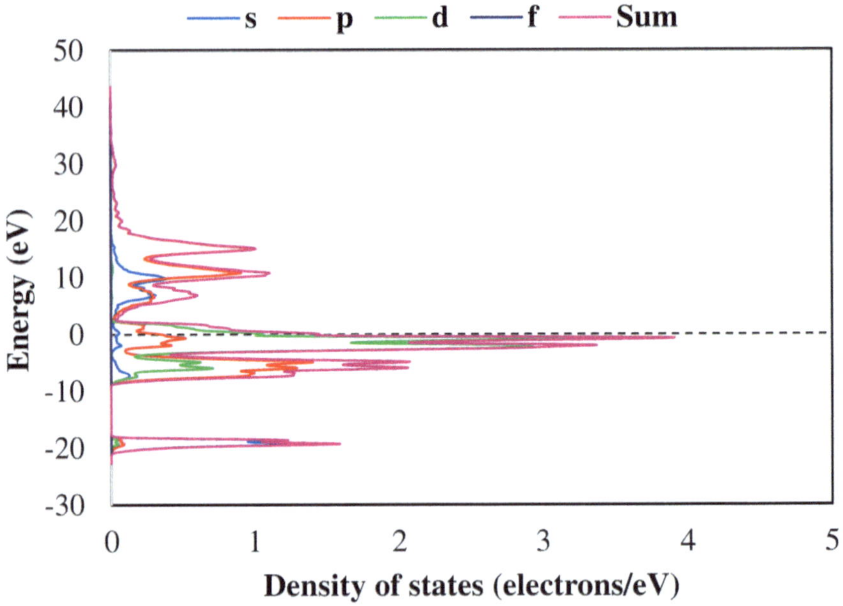

Fig. 2.15 **a** Electronic PDOS of NiO unit cell

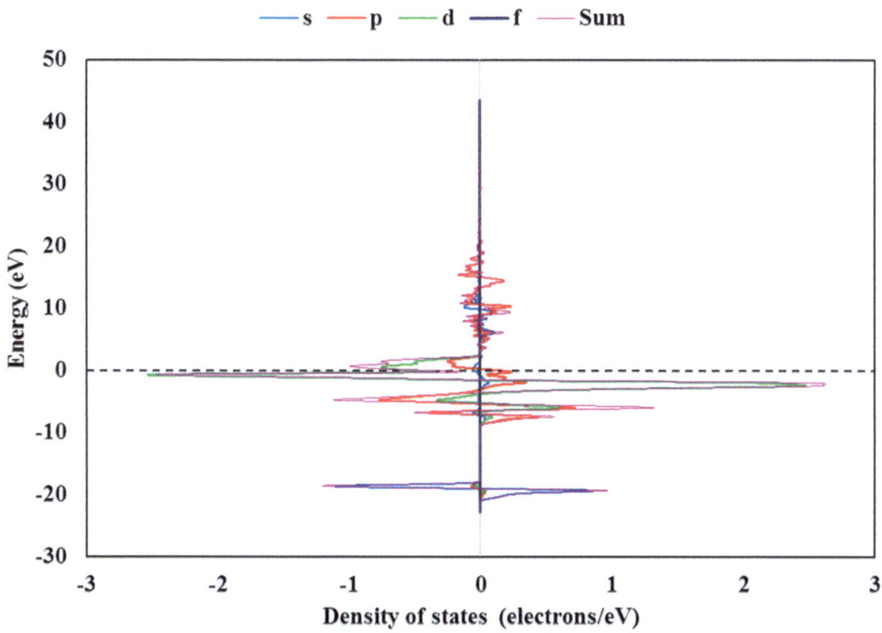

Fig. 2.16 Electronic net partial spin density of states (PSDOS)

maximum DOS 4 (electrons/eV). In 3d metal oxides, the catalytic and magnetic properties are strongly dependent on the d-center. Therefore, the maximum d-DOS lie between 3–4 eV within the energy range of 1 to −10 eV nearer below and slightly above the FL.

As the DFT calculations were performed for DOS for a spin-polarized system with a nonzero/formal initial spin, two charts automatically appeared as DOS-alpha (up spin) and DOS-beta (down spin). This setting enables us to calculate the net spin DOS of difference of alpha (up spin) and beta (down spin) DOS. Figure 2.16 illustrates the partial net spin density of states of subatomic orbitals (s, p, d, and total) of Ni and O in NiO unit cell structure. It can be clearly seen in Fig. 2.16 that the maximum SDOS lie between −3 and 3 (electrons/eV) 1 to−8 eV across the FL. Most of the SDOS lie above the FL between 0 and 20 eV, and a few below it between 0 and −10 eV. Most of the SDOS are attributed to p–d orbitals below and above the FL 2 to−8 eV. The maximum number of net spin of 2 is calculated.

2.4.2.2 DFT Simulation Study of Electronic Band Structure and Partial Density of States of NiO (111) Cleaved Surface

This section presents a simulation study based on the electronic band structure of NiO (111) cleaved surface as shown in Fig. 2.17. In Fig. 2.17, the blue–red lines in

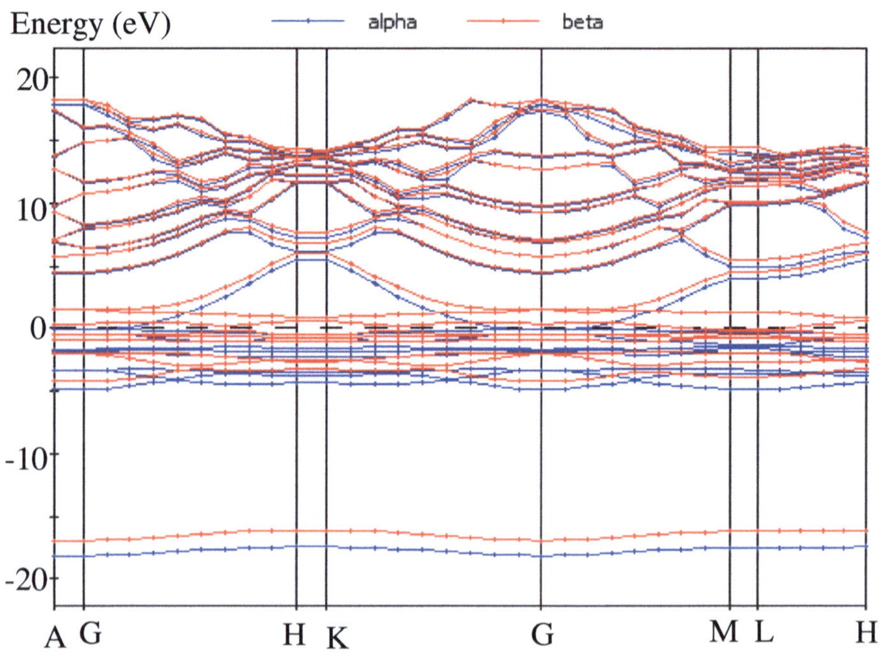

Fig. 2.17 Electronic band structure of NiO (111) unit cell

the electronic band structure of NiO (111) cleaved surface represent the electronic states of alpha (up) and beta (down) electrons for spin-polarized calculations across the Fermi level (FL). In comparison to the band structure of NiO unit cell, the electronic band structure is further split across the hove singularity/symmetry points/Brillion zone (G, A, H, K, G, M, L, and H) with narrow-populated electronic states with reduction in energy between −5 and −20 eV. This might be an indication of the quantum confinement effect in NiO because of transition from bulk NiO unit cell to NiO (111) cleaved surface. In Fig. 2.17, an evolution of narrow electronic states with minimum energy below and above the FL can be seen.

The corresponding PDOS of states of subatomic orbitals (s, p, d, and total) of Ni and O (Ni = 3d8, 4s2 and O = 2s2, 2p4) of NiO (111) cleaved surface are shown in Fig. 2.18 The maximum number of orbital density of states lies between 4 (electrons/ev) and 5 (electrons/eV) within the energy range of −1–5 eV. Moreover, mostly PDOS are contracted and populated between 20 eV and −5 eV above and below the FL. In d-orbital, there can be seen drop of energy between 3 and 5 eV and change of DOS near below 4 (electrons/eV) and above the FL.

The spin-polarized DOS of Ni and O in NiO (111) were found to be consistent with that of Ni and O in NiO unit cell as shown in Fig. 2.7. Again, it can according to Fig. 2.19 the maximum SDOS lie between −3 and 3 (electrons/eV) 1 to −8 eV across the FL. Most of the SDOS lie above the FL between 0 and 20 eV, and a few near below it between 0 and −10 eV. Most of the SDOS are attributed to p–d

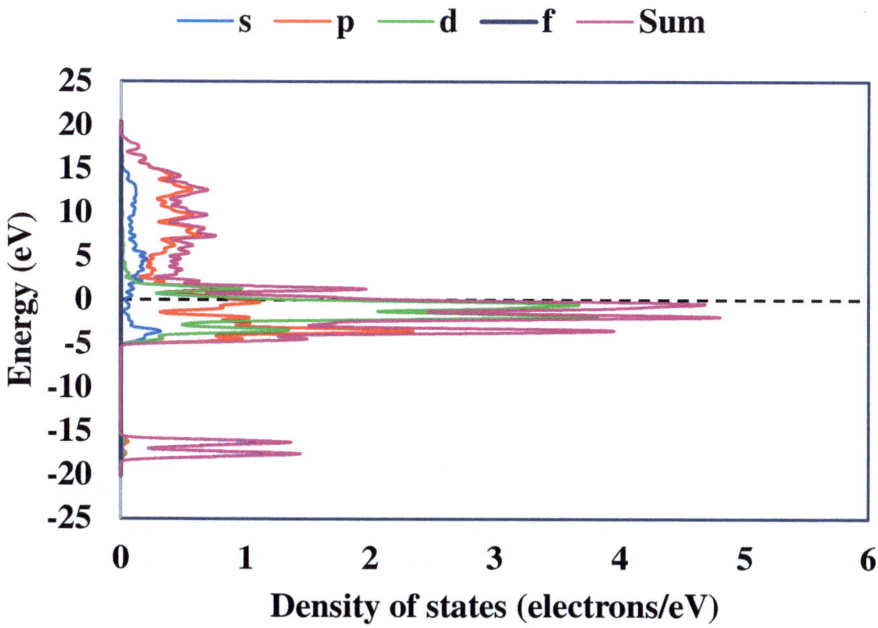

Fig. 2.18 Electronic PDOS of NiO (111)

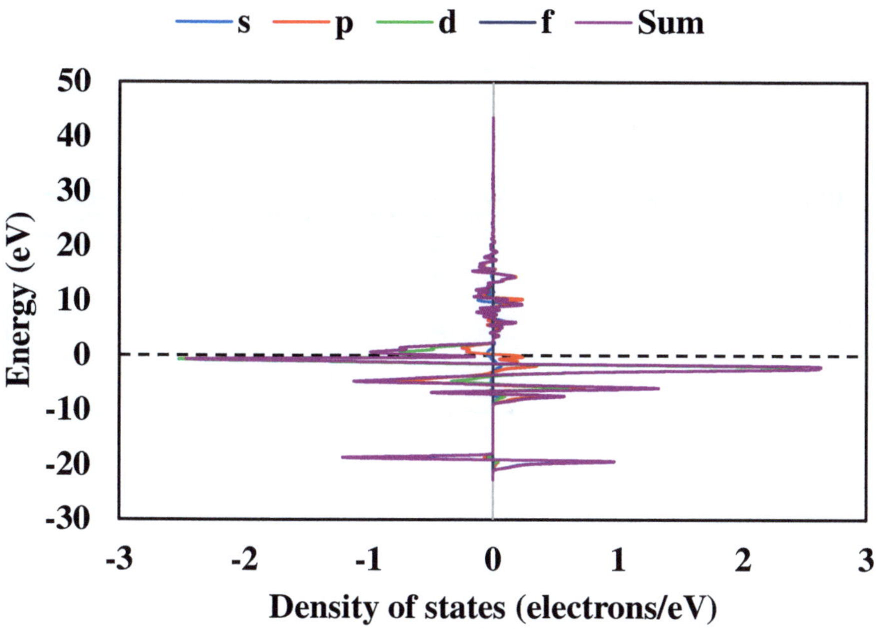

Fig. 2.19 Electronic net partial spin density of states (PSDOS) of NiO (111)

orbitals below and above the FL 2 to −8 eV. Ultimately, a maximum number of net spins of 2 is the outcome.

2.4.2.3 DFT Simulation Study of Electronic Band Structure and Partial Density of States of NiO (111) Cleaved Surface with CO_2

This section of the DFT simulation, investigates the effect of CO_2 molecule adsorbed over NiO (111) cleaved surface resulting in changes in the corresponding electronic band structure, DOS, PSDOS across the FL within conduction and valance band of sorbent NiO (111) nanocatalyst surface. The influence of the adsorption of CO_2 molecules is shown in Fig. 2.20 on the electronic band structure of NiO (111) cleaved surface at 7 Å. We observe a reduction or energy minimization of energy level of electronic states in the electronic band structure. The electronic band structure seems to be more dense or packed with closed split energy level with narrow band gap. It is quite vivid in Fig. 2.20 that conduction band of alpha–beta electrons above the FL overlaps with the valance band that is also

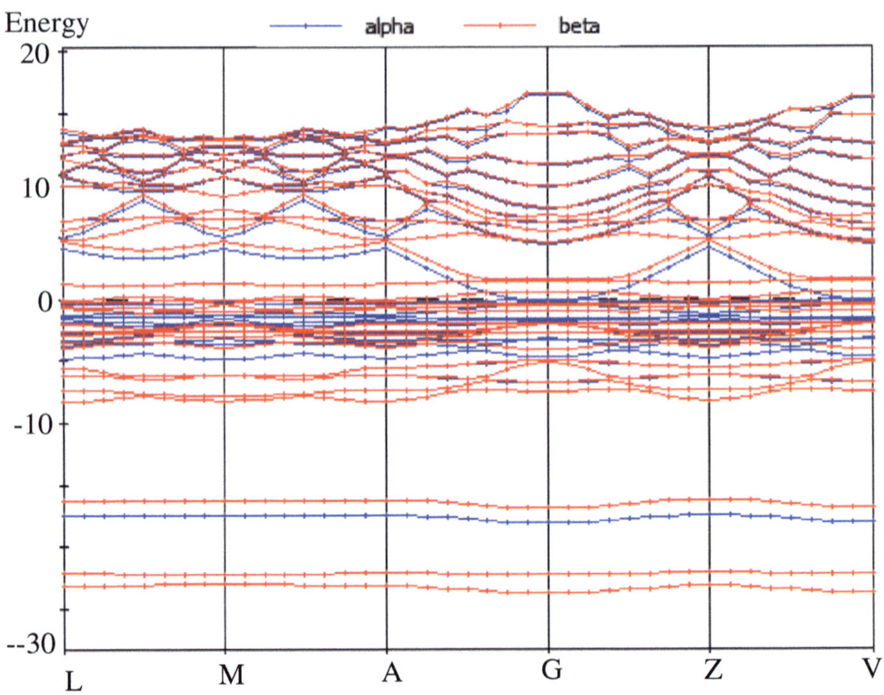

Fig. 2.20 Electronic band structure of NiO (111) with CO_2

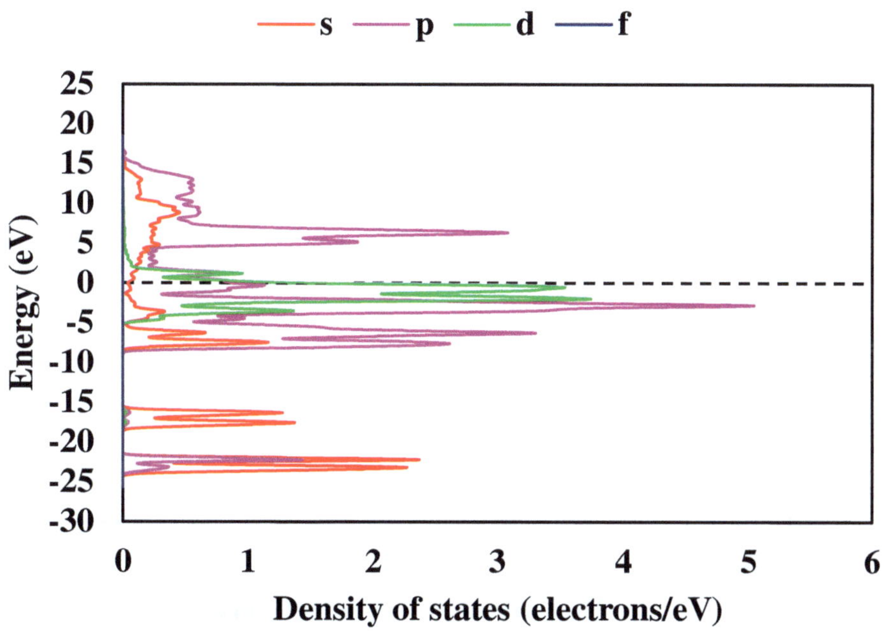

Fig. 2.21 Electronic PDOS of NiO (111) with CO_2

shifted slightly above and below the FL. This indicate the probable electronic transition of electrons from valance band to conduction band as well as their accumulation above the dangling NiO (111) cleaved surface for more electrostatic interaction and their wave function overlaps for appropriate catalytic interaction. The band structure tends to be more split across the hove singularity/symmetry points/Brillion zone (G, A, H, K, G, M, L, and H) with narrow populated electronic states and reduction in energy between 15 eV and –10 eV. An additional evolution of two red-(beta down up) is clearly observed in the valance band within the energy range of −20 to −25 eV.

The corresponding PDOS of subatomic orbitals (s, p, d, and total) of Ni and O (Ni = 3d8, 4s2 and O = 2s2, 2p4) of NiO (111) cleaved surface with CO_2 are shown in Fig. 2.21. The maximum number of orbital density of states lies between 5 (electrons/ev) and 6 (electrons/eV) within the energy range of −8 to 15 eV. The PDOS are increased and shifted further below between (−8 and 15 eV) because of the adsorption of the CO_2 molecules clearly indicating the orbital wave function overlapping and quantum mechanical perturbation among CO_2 and NiO (111). This causes a reduction in the band gap. The d-orbital DOS also seems to be modified and split for DOS around four (electrons/eV) and between 10 and −5 eV slightly above and below the FL.

With the adsorption of CO_2 molecule over NiO (111) cleaved surface, a slight modification of net spin partial DOS results, as shown in Fig. 2.22. This is attribute to the minor contribution to the enhancement of magnetic properties of the surface

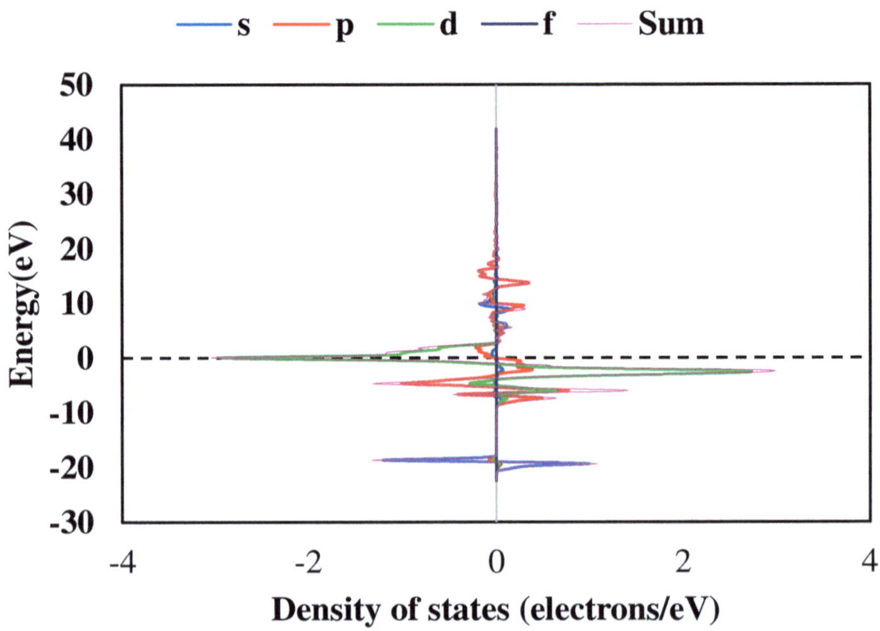

Fig. 2.22 Electronic net partial spin density of states (PSDOS) with CO_2

of NiO (111) cleaved surface. Figure 2.22, shows that total net SDOS has been increased from −3 (electrons/eV) to 3 (electrons/eV). Most of the net PSDOS lies below and above the FL with an energy range of 5 to −5 eV. The net spin DOS of d-orbital are also slightly modified across the FL. Partial net spin DOS analysis shows the slight modifications at each subatomic-level reactants CO_2 and nanocatalyst. Eventually, with the adsorption of the CO_2 molecule over NiO (111) cleaved surface, a modified net spin of three is the outcome.

2.4.2.4 DFT Simulation Study of Electronic Band Structure and Partial Density of States of NiO (111) Cleaved Surface with CO_2, H_2, and N_2

Figure 2.23 elaborates the electronic band structure analysis of NiO (111) cleaved surface with CO_2, H_2, and N_2 adsorbed. According to the electronic band structure analysis a further reduction of energy level is seen in the electronic states in the electronic band structure of alpha (up) and beta (down) electrons as compared to NiO (111) with CO_2. The of electronic states have been reduced between 12 eV above the FL to till −8 eV below the FL but with many up-down spin electronic states of sorbates and sorbent molecules within the valance level below the FL and within conduction band slightly above the FL. It clearly indicates a significant overlap of electronic states between valance band electrons and conduction band

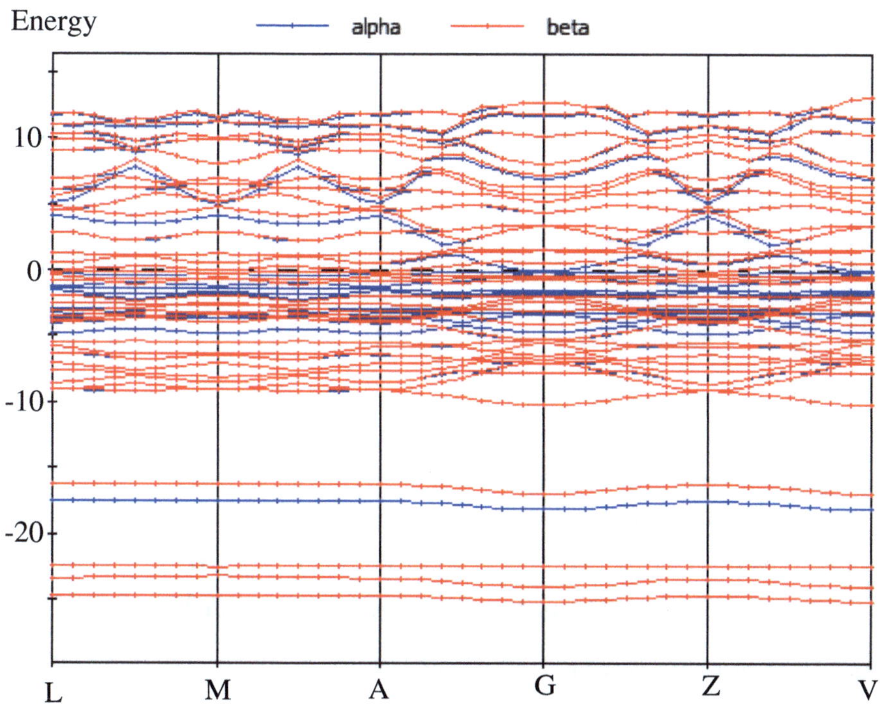

Fig. 2.23 Electronic band structure of NiO (111) with CO_2, N_2, and H_2

electrons and their parobable transformation from valance to conduction band. More charge/electron accumulation may be helpful above the nanocatalyst surface and more electrostatic interaction and their wave function overlap among reactant molecules and nanocatalyst NiO (111) for appropriate catalytic interaction. As previously mentioned, the band structure is seen to be more splitted across the hove singularity/symmetry points/Brillion zone (L, M, A, G, Z, and V) with narrow-populated electronic states reduced band gap and reduction in energy between 12 and −8 eV.

The corresponding PDOS of subatomic orbitals (s, p, d, and total) of Ni and O (Ni = 3d8, 4s2 and O = 2s2, 2p4) of NiO (111) cleaved surface with CO_2, H_2, and N_2 is illustrated in Fig. 2.24. The maximum number of orbital density of states lies between 7 (electrons/ev) and 8 (electrons/eV) within the energy range of −10 to 12 eV. The PDOS are increased and shifted further below between (−8 and 12 eV) because of the adsorption of CO_2, H_2, and N_2 molecule clearly indicating the significant orbital wave function overlapping and quantum mechanical perturbation among the reactant molecules (CO_2, H_2, and N_2) and NiO (111). This causes a further reduction in the band gap. The d-orbital DOS also seems to be modified and split slightly above and near below the FL. The ultimate PDOS are significantly modified.

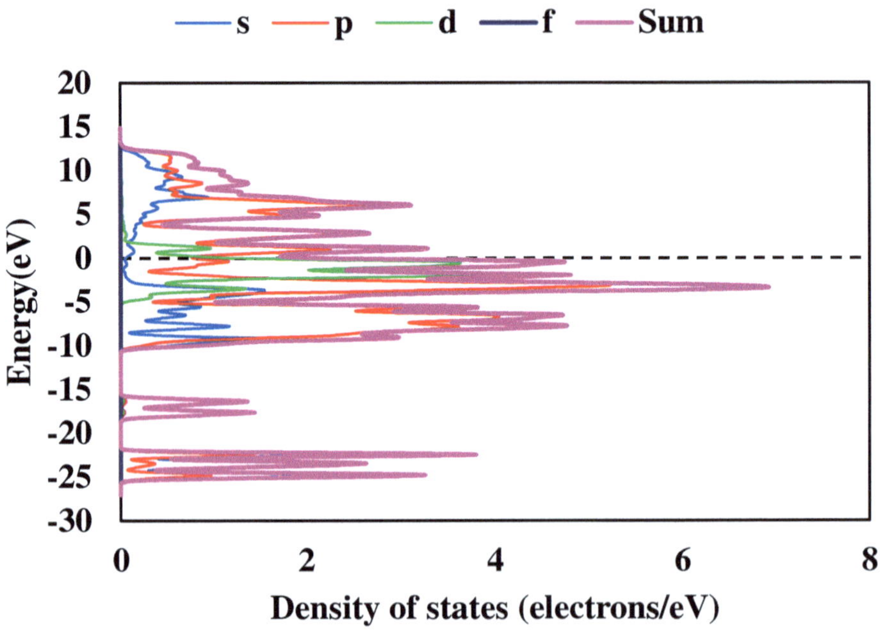

Fig. 2.24 Electronic PDOS of NiO (111) with CO_2, N_2, and H_2

The adsorption of three reactant molecules over NiO (111) cleaved surface attribute more toward the modification in net spin PDOS, magnetic properties of surface of NiO (111) cleaved surface as shown in Fig. 2.25. In fact this simultaneous adsorption of three reactant molecules results in a net spin of 5 instead of 2 and 3 for isolated NiO cubic unit cell, NiO (111) cleaved surface and NiO (111) with CO_2 adsorbed, respectively. This DFT simulation adsorption of reactant molecules also causes an increment in the number of electrons, number of bands, up spin (alpha), and down spin (beta) of sorbates and sorbent atoms. In Fig. 2.25, it can be seen that the total net SDOS was increased beyond −10 (electrons/eV) to 10 (electrons/eV). Most of the net PSDOS lie slightly below and above the FL within the energy range of 5 to −5 eV. Moreover, the d-orbital net SPDOS was significantly improved across the Fermi level, indicating a variation of slightly improved magnetic behavior of the nanocatalyst surface.

Table 2.2 clearly concludes that adsorption of CO_2 molecule over NiO (111) cleaved surface at a distance of 7 contributes to a significant effect over the electronic properties, band structure, DOS, and improved magnetic behavior as compared to the adsorption of N_2 and H_2 molecules.

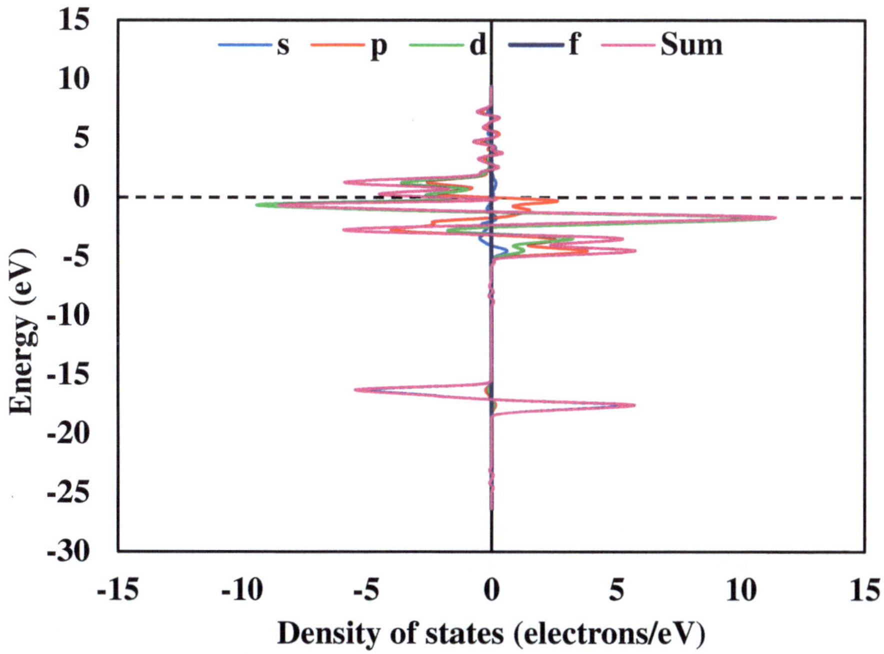

Fig. 2.25 Electronic net partial spin density of states (PSDOS) with CO_2, N_2, and H_2

Table 2.2 Electronic band structure, PDOS and net SPDOS of Ni and O in NiO unit cell, NiO (111), NiO (111), CO_2 and NiO (111) with CO_2, H_2, and N_2

Material	No. of electrons	Net spin	Up spin	Down spin	Subatomic orbitals	No. of bands
NiO unit cell	16	2	9	7	Ni = 2s2,3d8	9
NiO (111)	16	2	9	7	O = 2s2, 2p4	9
NiO(111), CO_2	32	3	17	14	H = 1s1	17
NiO(111), CO_2, H_2,N_2	23	5	23	18	N = 2s2, 2p3 C = 2s2,2p2	23
% difference NiO (111) with CO_2	50	33	47	50		47
% difference NiO (111) with CO_2,H_2,N_2	27	40	26	22		26

2.5 Nanocatalyst Synthesis and Characterization

Nickel oxide (NiO) nanoparticles (NPs) were synthesized by using a simple sol-gel technique. Initially, 5 g of Ni $(NO)_3.6H_2O$ (Sigma-Aldrich) was dissolved in 10 ml of deionized water. Next, 40 ml of ethylene glycol ($C_2H_6O_2$) (Sigma-Aldrich) was added to nitrate salt solution mixture at 60 °C under vigorous magnetic stirring of 350 rpm for 1 h. Furthermore, the solution mixture was cooled at ambient for xerogel formation.

This gel was dried at 90 °C for several hours. Finally, the greenish dry gel was burnt at around 400 °C for the formation of gray black nanocrystallites of NiO encapsulated by the core of Ni phase. The synthesized nanocatalyst was characterized for FESEM, EDX, TEM, Raman, FTIR, and VSM for size distribution, compositional analysis, phase transformation, and magnetic measurements.

2.5.1 Experimental Results of Synthesis and Characterization of Nanocatalyst

NiO nanoparticles (NPs) shells encapsulating the core nano Ni were synthesized through a simple sol-gel self-combustion route using an aqueous solution of nitrate salts and fuel following heating and rigorous magnetic stirring around 400 °C on a hot plate. The as-synthesized samples were eventually characterized for FESEM, EDX, TEM, Raman, FTIR, XRD, and VSM for morphological distribution, elemental analysis, phase transformation, and magnetic measurements.

XRD measurements of NPs are shown in Fig. 2.26. The prominent, strong, and sharp existing diffraction peaks of 2Θ at 37.3, 43.2, 62.8, 75.6, and 79.5 corresponding to diffraction planes (111), (200), (220), and (311), respectively, represent a pure phase of NiO nanocrystallite. Only a few diminished peaks of Ni phase of 2 Θ at 44.4, 51.8, and 76.4 corresponding to diffraction planes of (111), (200), and (220) were also observed. This vividly represents the formation of pure phase, cubic NiO (bunsenite, NaCl-type structure; space group; Fm-3 m; JCPDS# 98-001-9675),

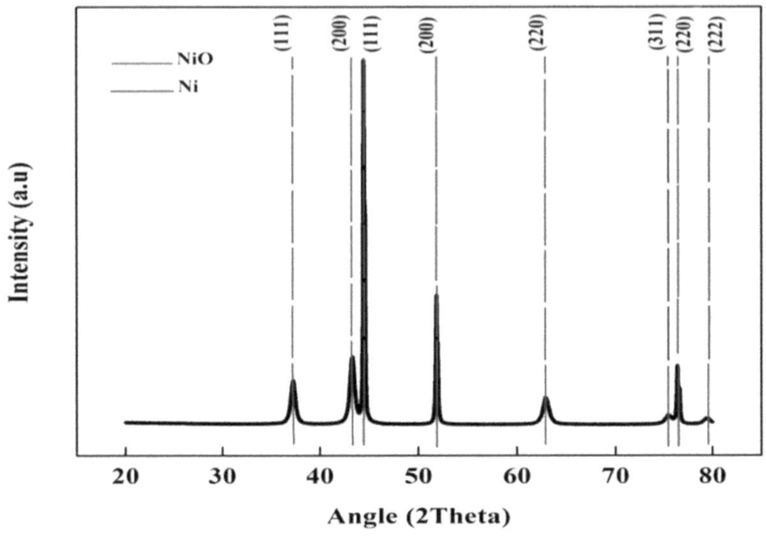

Fig. 2.26 X-ray diffraction analysis of core-Ni and shell-NiO NPs

and cubic Ni (space group; Fm-3 m, JCPDS# 98-002-2036) at this temperature. The average crystallite size and lattice strain of the synthesized nickel–nickel oxide phases of NPs were calculated from XRD data using Debye–Scherrer's equation

$$D_p = 0.9\lambda/\beta_{1//2}Cos\theta \quad (2.15)$$

where

D_p = average crystallite size
β = full width at half maximum
θ = Bragg's angle
λ = X-ray wavelength

is about 69.97 nm, 0.0013 and 56.2 nm, 0.003, respectively. The nature and positions of the above diffraction peaks are found to be the characteristic of cubic nickel–nickel oxide phase.

The FTIR and Raman spectroscopy results analyse is elaborate that Raman and FTIR shift (cm^{-1}) vs intensity are steady with dominant and broader peak at 510.17 cm^{-1} for NiO stretch phase Fig. 2.27b. Variation in intensity and peak shift (cm^{-1}) is also reported for as-synthesized nanomaterial. The primary objective to characterize the nanostructures by both characterization techniques was consistent in the presence of NiO phase at the oxide shell surface layer. The characterization results for peak intensities vs shift (cm^{-1}) by Raman and FTIR analysis with the presence of peaks between 490 cm^{-1} and 508.17 cm^{-1} for NiO stretching mode [2–6].

Three dominant Raman phases are eminent and positioned at 496.5 cm^{-1}, 510.17 cm^{-1}, 1076.66 cm^{-1}, and 1478.26 cm^{-1}, respectively [27, 28]. Considering the Raman shift change which could be inferred to strain/stress or size effects, the position and relative intensities of the dominant peaks are in good promise with the reported values of NiO phase matching to first-order transverse optical mode

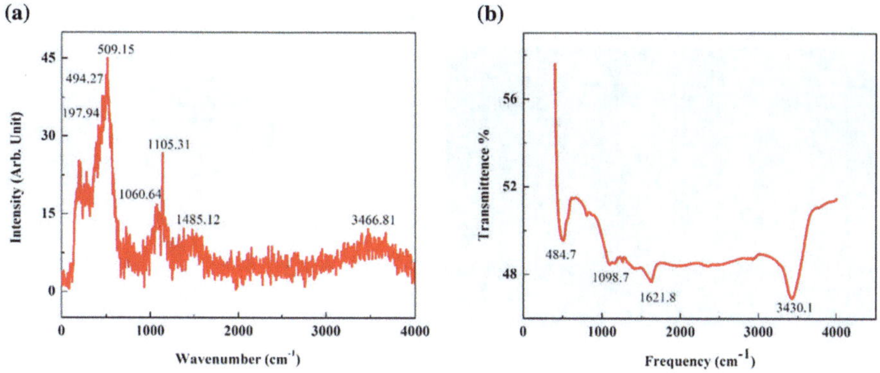

Fig. 2.27 a Raman spectroscopy b FTIR spectroscopy analysis of NiO NPs of as-synthesized material

(496.5 cm^{-1}), longitudinal optical mode (1093.1 cm^{-1}), and 2TOM (1524.8 cm^{-1}) phonon modes of vibration, respectively.

The dominant band at 1524.8 cm^{-1} is qualified to the scattering mode of the double magnon, and the small vibrational modes might be due to the existence of some impurities. In terms of FTIR, the dominant peaks for the NiO NPs observed are 474.7, 512.8, 818.6, 1098.74, 1417.9, 1621.8, and 3430.1 cm^{-1} between 400 and 4000 cm^{-1} [24, 27].

Peak values for 474.7 and 512.8 cm^{-1} attribute the presence of NiO phase in the powder sample, whereas the % T of 3430.1 cm^{-1} predicts the lower amount of water vapor. The less % T intensity for 34661.81 cm^{-1} peak reflects that most of the water content was elevated from the sample powder.

The average particle size of nanoparticles in Fig. 2.28a and b is between 5 and 13 nm by TEM analysis. Diffraction pattern confirms that the synthesized material is NiO FCC with prominent diffraction planes (200), (311), (400). In Fig. 2.28b, the dark black region surrounded by light dark one corresponds to Ni-core and shell-NiO respectively. Figure 2.28a represents an HRTEM diffraction pattern of partially oxidized as-synthesized NPs.

The image vividly exhibits the core-shell nature of synthesized nanoparticles. The core of the NPs is composed of Ni phase carrying single phase and of a relatively thin NiO outer shell made up of small nanocrystallites (polycrystalline) of NiO with diverse lattice alignments. In fact, during the self-combustion process around 400 °C in air, the oxidation might have taken place at the surface of NPs instead of penetrating the internal Ni nanophase.

In Fig. 2.29a, the FESEM analysis of NPs shows that NPs are spherical in morphology. Furthermore, the NPs seem to be agglomerated due to their ultrafine size distribution. The average size of as-synthesized NPs by autocombustion method around 400 °C is 17.6 nm which is by and large consistent as determined by TEM size distribution analysis. Figure 2.29b indicates the EDX elemental analysis of NPs

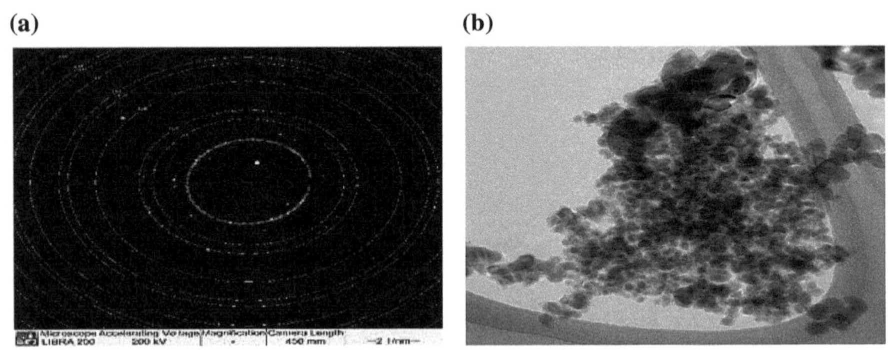

Fig. 2.28 TEM **a** diffraction pattern **b** size distribution analysis of as-synthesized NiO by autocombustion method

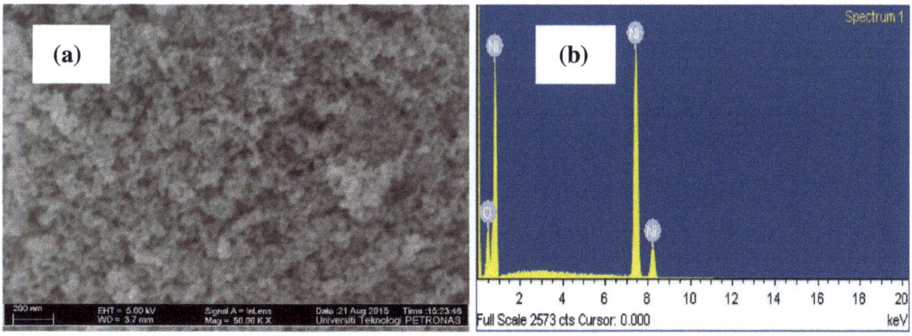

Fig. 2.29 **a** FESEM size distribution **b** EDX analysis of as-synthesized NiO NPs at 400°C

Table 2.3 EDX elemental composition analysis of NPs

Element	Weight %	Atomic
O K	9.20	27.10
Ni K	90.80	72.90
Totals	100.00	

with highest percentage of metallic-Ni contents rather than oxygen [3]. Table 2.3 presents the elemental composition details of NPs in term of weight % and atomic %.

The magnetic features of NPs were quantified by using VSM Lakeshore 7400 series. The entire hysteresis loop area value is 11.4×10^3 erg/g. The maximum value of saturation magnetization (M_s) of NPs is 38.2 emu/g (Fig. 2.30). The retentivity (Mr), coercivity (Hci), and squareness (Mr/Ms) values are found to be 4.43 emu/g, 98.7 G, and 0.12, respectively, for a maximum applied magnetic field of 20000 G.

The higher saturation magnetization values of as-synthesized NPs clearly indicate that the material carries a large core of Ni phase and surrounding thin shell of NiO phase [30]. As the synthesized NPs are quite smaller in size ($\sim$12 nm), the exchange bias effect might be less due to smaller core–shell volume ratio but still persists for NPs size distribution of 4 nm [13]. This feature of NPs causes less pinning or resistance (exchange bias) between antiferromagnetic (AFM) aligned Ni-core phase and disordered spin-glass (SG) state of NiO-shell [13, 14, 16].

These magnetic measurement results are quite consistent with XRD crystallite size and EDX elemental analysis of both Ni and NiO phases. The main objective to characterize these NPs for FESEM and VSM at various temperatures is to investigate the effect of their morphologies and saturation magnetizations on the output yield of green fertilizer.

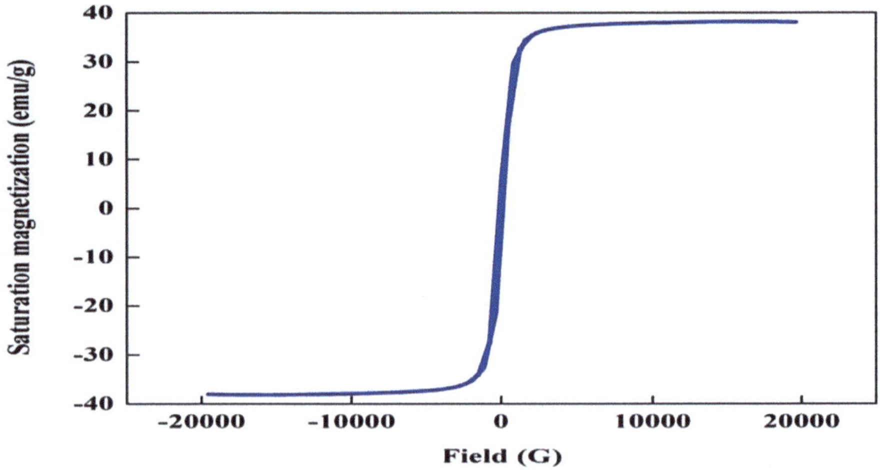

Fig. 2.30 VSM results of as-synthesized NPs at 400°C

2.6 Green Urea Synthesis Using Nanocatalyst in Magnetic Induction Method (MIM) Under Ambient Conditions

Green urea was synthesized under ambient conditions using NiO NPs carrying a core of Ni phase. H_2, N_2, and CO_2 gases were exposed inside a reactor with 0.75, 0.25, and 0.25 L/min flow rates, respectively. The mass of the NPs was 0.322 g, while the strength of the applied magnetic field was 0.2, 0.4, 0.5, 0.6, 0.8, 1.0, 1.2, 1.4, 1.5, 1.6, 1.8, 2.0, 2.2, 2.4, and 2.6 T. The duration of each experiment was restricted to one hour. The final output product (green urea) was diluted inside 10 ml of acetonitrile (ACN) solution. This accomplishment of series of experiments attributed to maximum yield of 6665.8 ppm at 2.4 T under ambient conditions.

Green urea synthesis under ambient conditions was carried out in microglass reactor as shown in Fig. 2.31. Core–shell NiO magnetic NPs were synthesized by self-combustion method and were used as nanocatalysts. These NPs containing a mass of 0.322 g were placed between the two poles of vibrating-sample magnetometer (VSM, Lakeshore-7400, and USA). Green urea reactant gases (N_2, H_2, CO_2) were exposed with a constant flow rate of 0.25 L/min, 0.75 L/min, and 0.25 L/min, respectively, with constant external applied magnetic field strengths of 0.2 T, 0.4 T, 0.5 T, 0.8 T, 1.0 T, 1.2 T, 1.4 T, 1.5 T, 1.6 T, 1.8 T, 2.0 T, 2.2 T, 2.4 T, and 2.6 T.

The outlet of the reactor was immersed into a glass cylinder containing 10 ml solution of acetonitrile (ACN) for the collection of desired fertilizer. Each experiment was conducted for 1 h. For each sample, the concentration of urea was quantified under calibration curves for 250, 750, 1250, and 1750 ppm incorporating with the experimental results of urea collected in ACN.

The maximum reported ammonia yield was 6665.58 ppm at 2.6 T applied magnetic field strength. The purpose to use NiO NPs as catalysts in this practice is

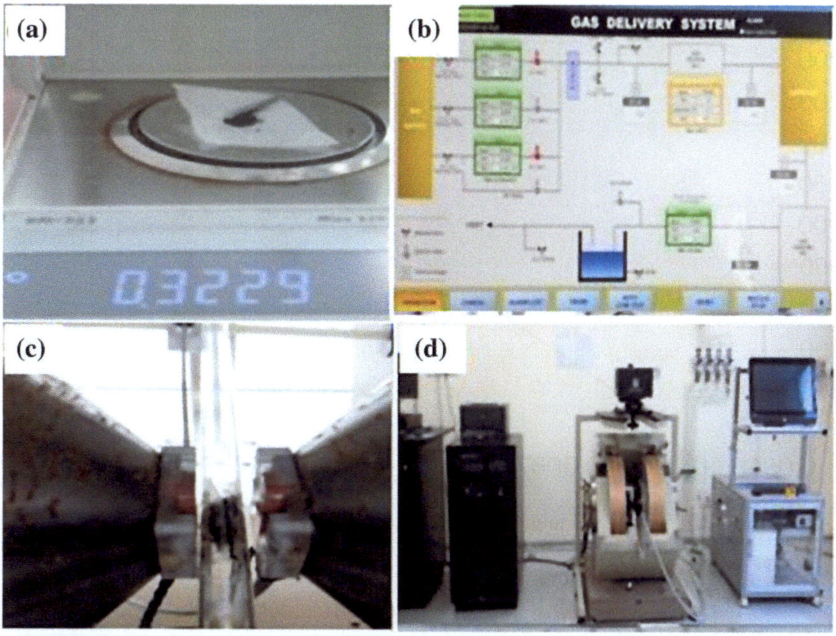

Fig. 2.31 Experimental setup for the synthesis of green urea in MIM **a** mass selection of NPs **b** electron micro flow controller of reactant gases display **c** microglass reactor **d** complete experimental assembly

because of their high surface area and wide range of applications in hydrogen sensing and hydrocarbon racking applications.

The calibration curve for urea vs absorbance values was developed by dissolving 250, 750, 1250, and 1750 ppm by dissolving 2.5, 7.5, 12.5, and 17.5 mg of commercial urea inside 25 ml of ACN. The FTIR results of pure ACN correspond to the minimum absorbance values in 1626–1670 cm^{-1} range. The FTIR results of commercial urea dissolved in ACN vividly elaborate the enhancement in the absorbance values of the solution mixture within 1600–1670 cm^{-1} and 3460–3700 cm^{-1} ranges for N-H bend and stretch modes with strong existence of urea between 1626 and 1670 cm^{-1} (Fig. 2.32).

The Fourier transform infrared spectroscopy (FTIR) experimental results in Fig. 2.33 show similar trends for increase in absorbance peak values and concentrations of green urea in ppm in the same range (1624–170 cm^{-1} and 3450–3700 cm^{-1}) as the results obtained from calibration curves by dissolving an increased concentration of commercial urea in 10 ml acetonitrile (ACN). In Fig. 2.33, FTIR results illustrate that with the increase in the external applied magnetic field strength the absorbance values gradually increase corresponding to improved urea concentrations. In practice, the experiments performed for 0.2 T–2.6 T, the maximum absorbance of 0.011 and urea yield was reported 6665.85 ppm at 2.4 T.

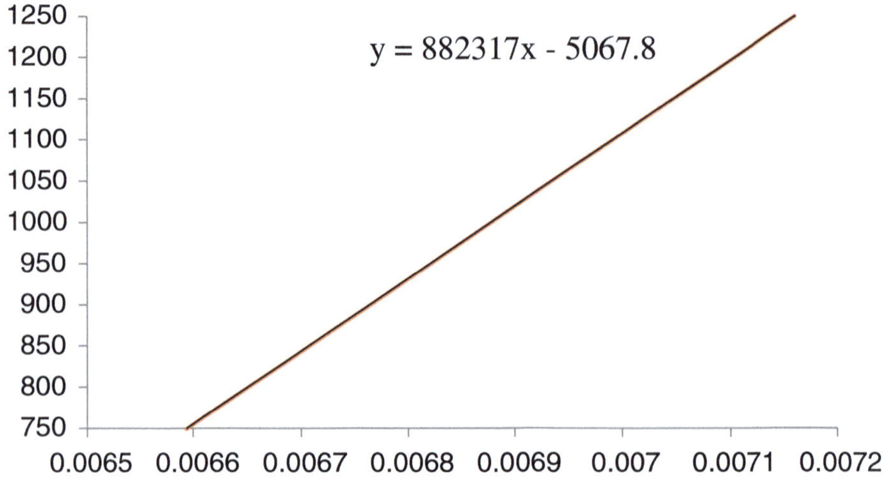

Fig. 2.32 Calibration curve for urea yield and absorbance from FTIR analysis

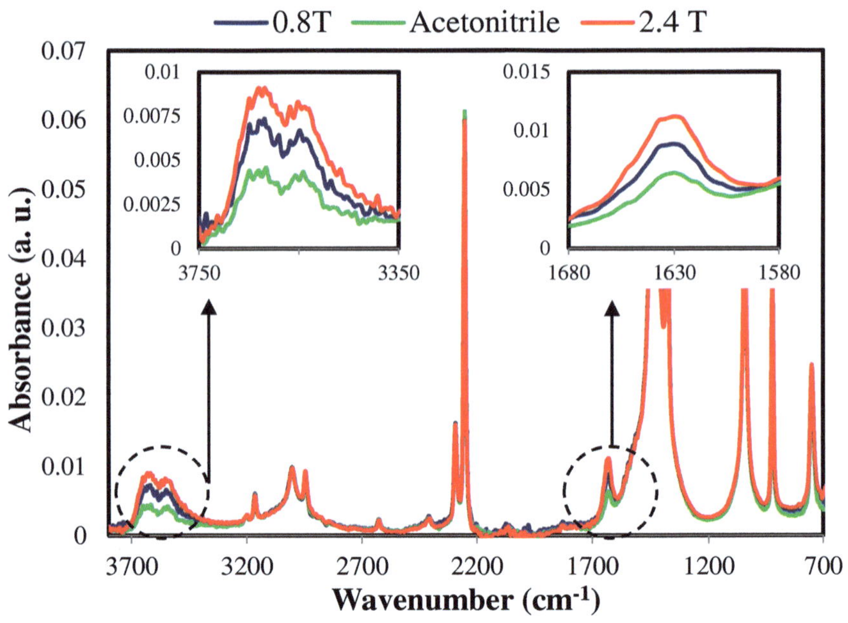

Fig. 2.33 FTIR over layer of ACN, green urea at 0.8 T to 2.4 T diluted in 25 ml of ACN between 1624 and 1667 cm^{-1} and 3300 to 3750 cm^{-1}

Table 2.4 presents the complete experimental details of green urea concentrations obtained for 0.2–2.6 T applied magnetic field strengths under ambient conditions for one hour.

Table 2.4 Green urea synthesis under magnetic induction method using core–shell NiO as catalyst

Applied magnetic field strength (T)	Urea yield (ppm)
0.2	1229.75
0.4	1277.41
0.6	2138.07
0.8	2343.30
1.0	2508.21
1.2	3343.81
1.4	3449.19
1.6	4001.25
1.8	4045.54
2.0	4490.36
2.2	4685.75
2.4	6665.85
2.6	4822.32

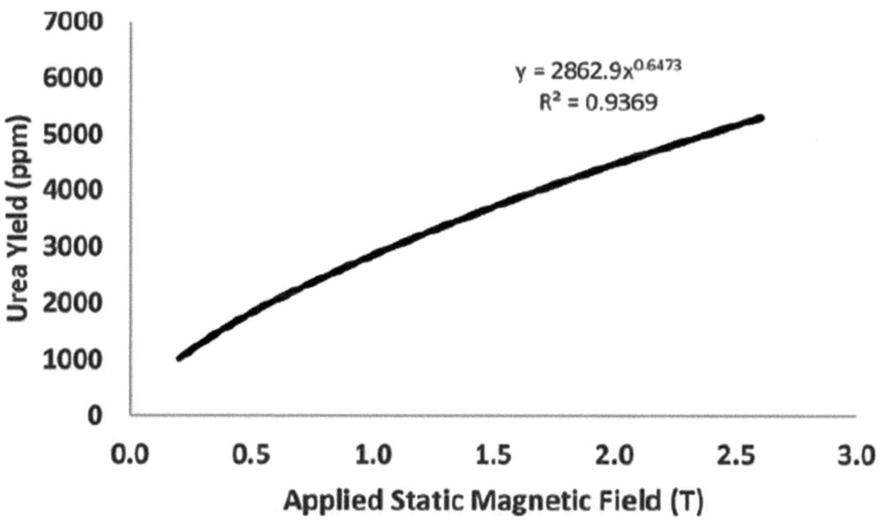

Fig. 2.34 Variation of green urea yield with static applied magnetic field strengths from 0.2 T to 2.6 T in MIM under ambient condition

The maximum green urea yield is 6665.85 ppm at 2.4 T applied magnetic field with a maximum absorbance in the FTIR graph within in the same range (1624–170 cm^{-1} and 3450–3700 cm^{-1}) as the results found from calibration curves by dissolving an increased concentration of commercial urea in 10 ml acetonitrile (ACN) (Fig. 2.34).

2.7 Summary

According to the CASTEP first principle DFT simulation results for F_2O_3, CuO, ZnO, NiO, the band gap PDOS, PSDOS were successfully generated. The simulation results elaborate the enhancement in the magnetic properties of both magnetic material metal oxides such as iron oxide and nickel oxide in terms of triplet (net spin) and magnetic moment modifications after the adsorption of reactant gases as compared to CuO (111), ZnO (101) with zero net spin or magnetic moment modification. CO2 adsorption over Fe2O3 (111) and NiO (111) attributes net spin and number of bands of 84%, 61% and 33%, 47%, respectively. The adsorption energy evaluation and triplet modification response of reactant over nanocatalysts are found to be slightly better for iron oxide than nickel oxide. The TEM, FESEM, EDX, Raman, FTIR, and VSM measurements vividly elaborate the experimental results of core-Ni and shell-NiO nanoparticles. The concentration of urea was quantified under calibration curves for 250, 750, 1250, and 1750 ppm incorporating with the experimental FTIR results of green urea in 10 ml of ACN in MIM. The maximum reported green urea yield was 6665.58 ppm at 2.4 T applied magnetic field strength.

References

1. Meessen, J. H., & Petersen, H. (2005). *"Urea", Ullmann's encyclopedia of industrial chemistry*. Weinheim: Wiley.
2. Austin, G. T. (1984). *Shreve's chemical process industries* (5th ed.). McGrawHill Inc.
3. Yahya, N., Qureshi, S., ur Rehman, Z., Alqasem, B., & Kait, C. F. (2017). Green urea synthesis catalyzed by hematite nanowires in magnetic field. *Journal of Magnetism and Magnetic Materials, 428*, 469–480.
4. Yahya, N., Puspitasari, P., & Noordin, N. H. (2013). Ammonia synthe-sis using magnetically induced reaction. In *Defect and diffusion forum* (Vols. 334–335, pp. 329–336).
5. Yahya, N., Alqasem, B., Qureshi, S., Irfaz, M., ur Rehman, Z., & Soleimani, H. (2016). The enhancement of the magnetic properties of α-Fe2O3 Nanocatalyst using an external magnetic filed of the production of green ammonia.
6. Yahya, N., Alqasem, B., Irfan, M., Qureshi, S., ur Rehman, Z., Shafie, A., & Soleimani, H. (2017). The effect of saturation magnetiza-tion of nanocatalyst and oscillating magnetic field for green urea synthesis. *Physica B, 507*,95–106.
7. Alnarabiji, M. S., Yahya, N., Hamid, S. B. A., Azizi, K., Kashif, M., Qureshi, S., & Alqasem, B. (2014). The role of surface area of ZnO nanoparticles as an agent for some chemical reaction. In *Defect and diffusion forum* (Vol. 354, pp. 201–213).
8. Hartley, H. (1919). Report of the British Mission appointed to visit enemy chemical factories in the occupied zone engaged in the production of munitions of war. In Dyestuffs: Hearings before the committee on ways and means, House of representatives, on H.R. 2706 and H.R. 6495, June 18–20 and July 14-18, 919 [66th Cong., 1st Sess., 1919], Committee on Ways and Means (Washington: Government Printing Office, 1919), 210–214.
9. Guillou, L. (2013). Urea-magnificent molecules. In *The Mole,* pp. 4.
10. MaxAppl. (1997). The Haber-Bosch Heritage: "The ammonia production technology". In *50th Anniversary of the IFA Technical Conference,* September 25–26th.

11. Flow Diagram of Urea Production Process from Ammonia and Carbon-dioxide, *Engineers Guide*, 2010.
12. Copplestone, J. C., Kirk, C. M., Death, S. L., Bitteridge, N. G., & Fellows, S. M. (2008) Ammonia and urea production, chemicals, ammonia and urea.
13. Brouwer, M. (2009). *Thermodynamics of the urea process*. Maastricht.
14. Kumar, B. & Das, P. C. (2007). Manufacture of urea, B. Tech.. National Institute of Technology Rourkela.
15. Brigden, K., & Stringer, R. (2000). *Ammonia and urea production: Incidents of ammonia release from the Profertil urea and ammonia facility, Bahia Blanca, Argentina*. Exeter: Greenpeace Research Laboratories.
16. Xiang, X., Guo, L., Wu, X., Ma, X., & Xia, Y. (2012). Urea formation from carbon dioxide and ammonia at atmospheric pressure. *Environmental Chemistry Letters, 10*, 295–300.
17. Kipriyanov, A. A., & Purtov, P. A. (2012). Magnetic field effects on chemical reactions near the disturbance of stationary states conditions. *Chaotic Modeling and Simulation, 1*, 53–65.
18. Steiner, U. E. (1983). Magnetic field effects on reaction yields in the solid state: an example from photosynthetic reaction centers. *Annual Review of Physical Chemistry, 34*, 389–417.
19. Ames, J. (2010). "Physical Chemistry", Chem 107B Lecture Notes. California: University of California, Davis.
20. Atkins, P., & De Paula, J. (2006). *Physical chemistry for the life sciences*. Oxford, UK: Oxford University Press.
21. Yamamoto, I., Ishikawa, K., Mizusaki, S., Shimazu, Y., Yamaguchi, M., Ishikawa, F., et al. (2002). Magneto-thermodynamic effects in chemical reactions. *Japanese Journal of Applied Physics, 41*(1, 1), 416–424.
22. Yamaguchi, M., Yamamoto, I., Ishikawa, K., & Mizusaki, S. (2002). Unified consideration on magneto-thermodynamic effects in chemical reactions. *RIKEN Review, 44*, 150–152.
23. Brown, T. L., LeMary, H. E., Bursten, B. E., Murphy, C. J., Woodward, P. M., & Stoltzfus, M. W. (2015). *Chemistry: The central science* (13th ed., pp. 668–698). Boston: Prentice Hall.
24. Everett, D. H. & Koopal, L. K. (2002). Chemisorption and physisorption. In Manual of symbols and terminology for physicochemical quantities and units. International union of pure and applied chemistry.
25. Davis, M. E., & Davis, R. J. (2003). *Fundamentals of chemical reaction engineering* (pp. 133–183). Mineola, New York: McGraw-Hill.
26. Reger, D. L., Goode, S. R., & Ball, D. W. (2010). *Chemistry: principles and practice* (pp. 542–548). Belmont, CA: Brooks/Cole, Cengage Learning.
27. Micromeritics Instrument Corporation, "Chemisorption", Micromeritics, USA.
28. Yahya, N., Puspitasari, P., Koziol, K., & Guiseppe, P. (2010). *"Ammonia Synthesis," in carbon and oxide nanostructures* (pp. 395–413). Berlin Heidelberg: Springer-Verlag.
29. Clark, S. J., Segall, M. D., Pick, C. J., Hasnip, P. J., Probert, M. I. J., Refson, K., et al. (2005). First principles methods using CASTEP. *Zeitschrift für Kristallographie, 220*, 567–570.
30. Clark, S. J., Segall, M. D., Pickard, C. J., Hasnip, P. J., Probert, M. I., Refson, K., et al. (2005). First principles methods using CASTEP. *Zeitschrift für Kristallographie-Crystalline Materials, 220*, 567–570.
31. Yahya, N., Irfan, M., Shafie, A., Soleimani, H., Alqasem, B., Rehman, Z. U., et al. (2016). DFT study of gases adsorption on sharp tip nano-catalysts surface for green fertilizer synthesis. In *AIP Conference Proceedings* (pp. 040011). AIP Publishing.
32. Dzade, N. Y., Roldan A., & de Leeuw, N. H. (2014). A density functional theory study of the adsorption of benzene on hematite (α-Fe_2O_3) surfaces. *Minerals, 4*(1), 89–115.
33. Verma, C., Olasunkanmi, L. O., Ebenso, E. E., Quraishi, M. A., & Obot, I. B. (2016). Adsorption behavior of glucosamine based pyrimidine-fused heterocycles as green corrosion inhibitors for mild steel: Experimental and theoretical studies. *The Journal of Physical Chemistry C*.
34. Harrison, W. A. (1989). *Electronic structure and the properties of solids*. Dover Publications.

35. Scheffler, M., Stampfl, C. (2000). Theory of adsorption on metal substrates. In *Handbook of surface science.* Berlin.
36. Vázquez, M. (2015). Magnetic nano-and microwires: design, synthesis, properties and applications. Woodhead Publishing.
37. Qureshi, S., Yahya, N., Kait, C. F., Alqasem, B., ur Rehman, Z. Irfan, M. (2017). Enhanced catalytic activity of α-Fe2O3 with the adsorption of gases for ammonia synthesi. *Materials Science Forum, 880*, 15–18.

Chapter 3
Reactor Design for Novel Green Urea Synthesis

3.1 Introduction

Fertilizer is considered a kingpin in enhancing the overall production of crops. Fertilizers, whether from inorganic or organic natural sources, have been persistently applied to enhance or promote crop growth in order to satisfy the increasing worldwide demand. An improvement in fertilizer efficiency in terms of nutrient uptake and crop yield is important for fertilizer both producers and users. However, the advert effects of the utilization of fertilizers on the environment have been a major discussion particularly by environmentalist groups. Improper and extensive utilization are among the major concerns that need to be addressed. One needs to take a step back on the production rate, and further theoretical evaluation must be carried out for better insights on the feasibility of the process. Subsequently, a release pattern can be evaluated for better understanding of the nitrogen-based fertilizer.

Ammonia-based fertilizers have been considered a key factor in developing and promoting agriculture industry in Malaysia [1]. The heart of the process of fertilizer production is the ammonia synthesis reactor. Typical Haber–Bosch method of ammonia synthesis, which was patterned in 1916, requires high energy input of 500 °C temperature level and 300 bar pressure. The current pressure of high energy costs motivates research to develop greener and low energy-intensive processes for ammonia production. In the last century, a great deal of research was carried out to improve the ammonia yield particularly those related to catalysis studies [2–4]. This project investigates a new synthesis approach employing magnetic inductance at ambient temperature and atmospheric pressure [5]. Urea synthesis from ammonia and carbon dioxide was discovered at the end of the nineteenth century, as summarized in the following equations.

$$N_2 + 3H_2 \leftrightarrow 2NH_3 \tag{3.1}$$

N. Yahya, *Green Urea*, Green Energy and Technology,
https://doi.org/10.1007/978-981-10-7578-0_3

$$2NH_3 + CO_2 \leftrightarrow NH_2COONH_4\ (\textbf{ammonium carbamate}) \tag{3.2}$$

$$NH_2COONH_4 \leftrightarrow H_2O + NH_2CONH_2 \tag{3.3}$$

The current method for the progress of conventional urea synthesis of reaction for this equation in terms of reaction enthalpy is shown in Fig. 3.1a. A one-step method of urea synthesis in the magnetic induction reaction zone is proposed as shown in Fig. 3.1b. The total energy consumption is typically between 30–35 Gj/ton NH_3, mostly attributed to the high demand of energy for compression and heating. Yahya and Puspitasari [5] successfully produced ammonia at ambient pressure and temperature giving 46% ammonia yield as compared to 14% ammonia yield from current industrial processes. The group employed the synergetic effect of magnon–phonon interaction during the reaction in the presence of magnetic induction (PI 2010005696). While the new magnetic induction reaction approach offers tremendous potential for energy savings, it lacks fundamental understanding of the mechanism of magnetic induction to effect the reaction equilibrium.

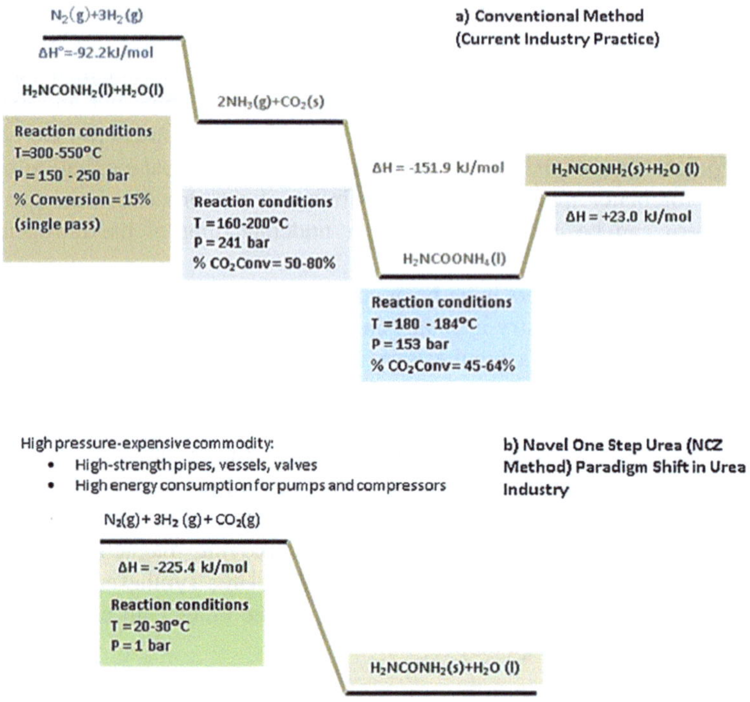

Fig. 3.1 Ammonia–urea synthesis **a** conventional and **b** novel one-step method

For any given chemical reaction, the change in standard Gibbs free energy, ΔG° of the system allows the determination of whether the system is spontaneous. The relationship between ΔG° and equilibrium constant K is given as:

$$\Delta G^{\circ} = -RT \ln K$$

where

ΔG° = Change in Gibbs free energy
R = Gas constant
T = Temperature
K = Equilibrium constant

Spontaneity occurs when $\boldsymbol{\Delta G^{\circ}}$ is negative when the equilibrium constant, K > 1, indicating a forward reaction is favored, thus enriching the ammonia or urea production. Therefore, $\boldsymbol{\Delta G^{\circ}}$ for a particular reaction is large and negative; it would be expected for the reaction to go to near completion. For ammonia synthesis, $\boldsymbol{\Delta G^{\circ}}$ is −33.32 kJ compared to −197.2 kJ for urea synthesis from a mixture of N_2, H_2, and CO_2 gases. The larger negative value of $\boldsymbol{\Delta G^{\circ}}$ for urea synthesis that allows researchers to predict the urea production is more spontaneous as compared to ammonia synthesis.

The concept of energy, based on the second law of thermodynamic, provides a thermodynamic characteristic that describes the potential of any form of energy to perform useful work with respect to the state in the environment. It is proposed that for ammonia to be formed, work is needed to drive the reaction equilibrium toward product. Exergy of a substance can be separated into physical and chemical [6, 7]. While physical exergy has been extensively explored [8], not much is understood on the concept of chemical exergy. The key to understanding reaction equilibrium is the Gibbs free energy. It is interesting to note that energy load expressed as Gibbs free energy is similar to chemical exergy with reference to the environment condition of temperature $\boldsymbol{T_{\circ}}$.

$$\Delta Ex = \Delta H - T_{\circ}\Delta S$$

Additionally, chemical exergy level may be related to the environment condition by definition of chemical potential

$$\mu = -\Delta G + RT \ln(P/P_o)$$

A fundamental understanding on the role of chemical potential μ and its interaction with the load of reacting species is to be explored. The interaction between μ and $\boldsymbol{\Delta Ex}$ is investigated so that the minimum level of chemical potential of reactants that would affect the equilibrium reaction on the forward direction could be determined.

Iron metal with five unpaired electrons as nanocatalysts are the main focus in this work. The catalytic activities of iron are intimately correlated to the mechanism of catalysis on the active centers. The interaction between homogenous magnetic field in the vicinity of active centers and the magnetic moments of the gaseous particles, in this case one mole of nitrogen and three moles of hydrogen, should be explicitly done to reflect the specific nature of the catalytic reaction. Based on similar principles, it is also anticipated that one-step urea production from nitrogen, hydrogen, and carbon dioxide gases could be conducted at ambient conditions (Fig. 3.1b). A one-step NCZ approach on urea synthesis employing nitrogen gas and synthesis gas as the feedstock is conducted in a high-density magnetic flux with the new catalyst support system which is assisted by rigorous CFD simulation. The proposed one-step NCZ urea synthesis system is depicted in Fig. 3.2. The microreactor design under patent filing (PI 2011001512) employs a magnetic reaction in ambient condition.

We propose one-step urea synthesis whereby three reactants (N_2, H_2, and CO_2) come in contact with the catalyst in the magnetic induction reaction zone, thus resulting in the formation of aqueous urea. The proposed process occurs at ambient temperature and atmospheric pressure, which effects savings in a variety of ways:

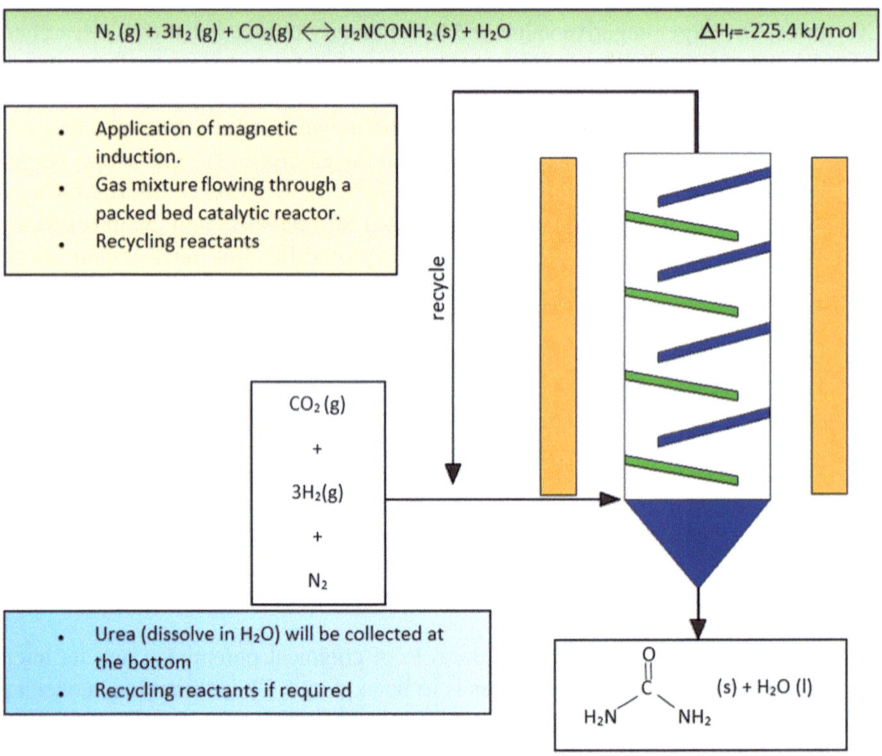

Fig. 3.2 Proposed one-step NCZ microreactor system

- As there is no need to pressurize gases at a high pressure, compressor work is saved.

  This has been calculated as 150 kWh/ton ammonia.

- There are savings in the costs of compressor/drive motor/low-pressure reactor.
- Energy is saved in the recycling of unreacted gases back to the reactor.

On account of the low-pressure of operation, the reactor does not require high-cost materials that are capable of operation at high temperatures. There are also savings in the fabrication of the reactor and other components that work at lower temperatures. All of these led to considerable savings in energy and associated costs.

The technique of replacing thermal energy with magnetic energy leads to a new approach for driving the force of chemical reactions. Since magnetic flux is used here as opposed to temperature, a new form of the Arrhenius equation is required. This new era of research may be termed as thermos-magnetic induction equilibrium reaction. As comparison is required between heat, work (as mechanical energy), and magnetic energy, this gives rise to a novel approach based on the second law analysis.

Process simulation plays an crucial role in enabling quantitative assessment and evaluation of various process structures to be accomplished without the need to construct a pilot plant scale for the new process. The novel thermodynamic models developed from the experimental data obtained are incorporated into a commercial simulator for simulation study to be conducted for optimization and overall feasibility testing of the process. In addition, a benchmarking study will also be undertaken to compare the performance of the new process against its conventional Haber–Bosch counterpart.

In urea synthesis, atomization is an initial step which must be experienced by ammonia and carbon dioxide molecules, prior to proceeding to the next steps of urea formation. The adsorption process allows molecules to dissociate, through physisorption and chemisorption eventually forming monoatomic. Rearrangements of these molecules take place for the formation of urea molecules prior to the desorption process. This part of work was evaluated in Chap. 2 of this book.

Capital, feedstock, and logistic escalating prices have altered business decisions. Moreover, the need to support zero-emission for sustainable future (Chap. 1) growth is continues due to the increase in demand of the N type fertilizer. In order to overcome this drawback, magnetic nanotechnology is potentially an excellent solution. The catalyst which is in a magnetic phase is able to reduce the activation energy; hence, an increase in the overall yield can be achieved. It is well-known that nanocatalyst with high saturation magnetization and with a surface with sharp tips (nanowires) will significantly improve the yield of urea. The sharp tips are occupied by high electron and spin density which lead to greater induced magnetic fields [6]. The adsorption of carbon dioxide gas on the surface of the magnetic catalyst was able to improve the magnetic properties of the catalysts. This offers a great deal of potentials to validate the urea by testing this new synthesis route using magnetically induced reaction. Scientists have shown special concern in studying the effect of magnetic and electromagnetic (EM) fields on the rate of chemical reactions and the related processes [7–11]. The yield of chemical reactions is greatly influenced by the magnetic field [12, 13]. This influence is due to the conversion of singlet to triplet states of reactants caused by the modification of their reaction rates [13–15]. The rate of chemical reaction and extent of spin mixing can be altered either by energetically separating the triplet sublevels via Zeeman interaction with static magnetic fields of typically 1 mT and above by improving the mixing process through the application of low static magnetic fields, or by the electric fields which play a vital role in charge transfer among adsorbates and adsorbents that further reinforce the energy level splitting at subatomic levels for triplet transformation [16, 17]. The effect of the oscillating magnetic field on the chemical reaction, theoretically and experimentally, performed at frequencies corresponding to hyperfine splitting shows a change in the yield of the chemical reactions [18–20]. The coupling of applied electromagnetic waves and static magnetic field activates the nanocatalysts which induce Zeeman, hyperfine, and stark splitting at their subatomic and molecular energy levels.

A novel method was employed to synthesis ammonia and urea at lower heat energy and lower pressure configurations. This can only be made possible when nanocatalysts are under the effect of a magnetic field which promises higher yield and improved safety [21–25]. Applying magnetic field on the nanocatalyst and reactant gases causes the percentage conversion from singlets to triplets to increase, thus leading to an increase urea yield [26–28], whereas the electric field component from the electromagnetic field executes electric forces on the electron distribution of reactant molecules. Consequently, this causes Stark effect splitting to occur, as well as the generation of appropriate polarization that enhances the catalytic activity of the nanocatalyst which improves the green urea yields [29, 30]. It should be noted that green ammonia synthesis is also based on these fundamentals.

3.1.1 The Effect of Magnetic Field on Singlet to Triplet Conversion

Chemical reactions can be affected by magnetic fields which can change the yield, rate, or output distribution [31]. Chemical reactions can occur without change of spin in the electron-pairs or with uncoupling of the electrons. In the former case, the spins in the electron-pairs remain balanced, chemically termed as heterolytic; in the latter case, unbalanced individual electron spins appear to be classified as homolytic or radical. The intrinsic angular momentum of both electrons and nuclei is conserved in elementary chemical reactions involving substitution, dissociation, recombination, electron transfer, addition, disproportionation, as well as many others (i.e., these elementary processes occur without changing the spins of electrons and nuclei). This conclusion applies equally to homolytic and heterolytic reactions, as directly indicated by the results of the chemically-induced magnetic polarisation of electrons and nuclei. The first chemical magnetic field effects

(MFEs) were observed in the sixties by Johnson, and the triplet-exciton annihilation luminescence of anthracene crystals. Brocklehurst mentioned that the output of recombination reactions between radical ions would show MFEs. The radical pair mechanisms supply the mechanistic basis for the translation of most magnetic field effects in chemical systems. To date, the effects of magnetic field that are known vary significantly, from magnetic field strengths ranging from about 40 μT to some Tesla [32].

Le Chatelier's principle states that a change in temperature, pressure, or concentration of reactants in an equilibrated system stimulate a response. Changes in temperature by adding or removing heat specifically shift the equilibrium [33]. Figure 3.1 demonstrates the effect of equilibrium conversion on both exothermic and endothermic reactions when the temperature is modified. Equilibrium conversion is the highest conversion that can be achieved by the reaction [34]. As the temperature increases, the equilibrium conversion X_{Ae} increases for endothermic reaction (absorb heat to produce the product). While in exothermic reaction (release heat to form the product), the equilibrium conversion decreases as temperature increases [33].

Endothermic reaction is when the sum of enthalpy in reactants is larger than the enthalpy of product itself. The energy absorbed to overcome the force of attraction between reactants molecules is higher than the energy absorbed to form new bonds. In contrast, in an exothermic reaction, enthalpy of product is greater than reactant. The energy to overcome the molecular force of reactant is lower than the energy to bond the product itself [33].

The change in equilibrium conversion affects the production of final product (urea). As temperature increases, almost all reactants in the endothermic process, which are ammonia and carbon dioxide are converted to urea. Conversely the amount of product (urea) formed is small for exothermic process when temperature increases [34].

K is equilibrium expression that can be written as [35]

$$aA + bB \rightarrow cC + dD \tag{3.4}$$

$$K = \frac{[C]^c [D]^d}{[A]^a [B]^b} \tag{3.5}$$

where *a*, *b*, *c*, and d are stoichiometric coefficient.

The relation of equilibrium conversion X_{Ae} with equilibrium expression is as follows [3/36]:

$$X_{Ae} = \frac{K}{K+1} \tag{3.6}$$

3.1.1.1 Magneto-Thermodynamic Effect in Chemical Reaction

This subsection is concerned with the magnetic field effects (MFEs) on the chemical equilibrium and related phenomena. Besides the equilibrium constant, a chemical reaction is accompanied by important thermodynamic quantities, such as the heat, free energy, and entropy of the reaction. Consider the chemical system in which k kinds of ideal gases are reacting with each other at the stoichiometric coefficients $v_i(i = 1 \sim k)$ under the influence of a magnetic field. The condition of a chemical equilibrium is written as follows:

$$\sum_{i=1}^{k} v_i \mu_i = 0 \tag{3.7}$$

where the chemical potential of the ith gas consists of the non-magnetic term $\mu^{(0)}i$ and the magnetic term $\mu^{(m)}i$ as

$$\mu_i = \mu_i^{(0)} + \mu_i^{(m)} = \mu_i^* + RTln\frac{P_i}{P*} - \int_0^H \mu_0 m_i dH \tag{3.8}$$

where μ_i^* is the standard chemical potential at temperature T and the standard pressure P^* in a zero magnetic field. P_i and m_i represent the partial pressure and the molar magnetization for the ith component. μ_0 is the permeability of the vacuum. The integration must be done with respect to the magnetic field strength H at constant temperature T. The equilibrium constant K_p is derived from Eqs. (3.2) and (3.3) as

$$\ln K_p = \sum_{i=1}^{k} v_i \ln\frac{P_i}{P*} = -\frac{1}{RT}\sum_{i=1}^{k} v_i \left(\mu_i^* + \mu_i^{(m)}\right) \tag{3.9}$$

Equation (3.4) leads to the magnetic field-induced change in the equilibrium constant

$$\ln K_p^{[H]} - \ln K_p^{[0]} = -\frac{1}{RT}(-\sum_{i=1}^{k} v_i \int_0^H \mu_0 m_i dH) = -\frac{1}{RT} g^{(m)} \tag{3.10}$$

where $^{[H]}$ and $^{[0]}$ represent the equilibrium constant in the magnetic field and in the zero fields, respectively. Similar notations are used for other thermodynamic quantities below. $g^{(m)}$ is the magnetic free energy change per unit reaction. Moreover, the heat of reaction ΔH^0, the free energy of reaction ΔG^0, and the entropy of reaction ΔS^0 are a function of temperature T and magnetic field strength H, as well. In this paper, we omit the $^{[0]}$ form the notation for simplification. Using the standard theory of chemical thermodynamics yields the magnetic field-induced changes in these thermodynamics quantities as

$$\Delta H^{[H]} - \Delta H^{[0]} = g^{(m)} - T\frac{d}{dt}g^{(m)} \tag{3.11}$$

$$\Delta G^{[H]} - \Delta G^{[0]} = g^{(m)} \tag{3.12}$$

$$\Delta S^{[H]} - \Delta S^{[0]} = -\frac{d}{dt}g^{(m)} \tag{3.13}$$

Equations (3.5)–(3.8) are the general formulae for the magneto-thermodynamic effects in chemical reactions. This magneto-thermodynamic formulation is applicable to all of the MFEs in chemical reactions [36].

3.1.2 The Development of Reactor

Technology advancement is key to sustainable future. Significant development in microelectronics has been vast over the past few years. Remarkable achievement has been reported which resulted in greater overall quality and efficiency. In the 1970s and 1980s, the chemical process industries emphasized more on the economical scale [26]. For the petrochemicals industries, this became the core motivation to increase the yield and production rate, rates of nearly 2,000 tons/day. Similarly, the chemical industry acknowledged the phrase "big is beautiful"—which no longer only applied to the oil and gas industry. For this reasons, the focal confrontation by the chemical industry could cover plant security, efficiency and ecological protection, temperature control, and efficient mixture in large batch vessels [26].

A major consideration today is increasingly being given to the economical scale by involving intensification processes and microreactor technologies. The target of expanding these methods is to achieve process plants that are run more efficiently with higher yields, less energy consumption, smaller footprints, improved safety, and localized approaches compared to conventional plants. Although microreactor technology is an extraordinary subject, it is a practical benchmarking against the conventional approach. Intrinsically, microreactors are small in size and have large surface-to-volume ratios that make massive heat transport possible. There are more significant advantages offered by microreactor compared to conventional large-sale reactors [26]. It is worthy to note that this would be the fundamental understanding before scale-up strategies are formulated.

3.1.2.1 Phase 1: Y-Shape Micro-reactor

Within the past two decades, microreactor technology has evolved from a highly advanced toy for chemical engineers to a versatile tool for chemical synthesis. Previously, the only way to conduct solution-phase synthesis was the conventional batch mode in stationary reactors with stirring or shaking as the only means to mix reactants. Today, microstructured devices offer significantly enhanced mixing and heating capabilities compared to the batch process, leading to improved product profiles and higher yields. There are numerous benefits that come with microreactor technology such as scale-independent synthesis, product profile improvement, accelerated process development, enhanced safety, constant product output quality, cleaner product profile, as well as higher yields.

Organic chemists typically spend a substantial amount of time on the development of synthetic routes to new materials. The best pathways need to be chosen, and reaction conditions must be optimized. Once a product is successful and demand for larger quantities grows, the entire synthesis process needs to be revised and readjusted for larger batch sizes. This general procedure shows a typical problem in conventional batch synthesis since batch synthesis is a space-resolved process. The output of the reaction is determined by the size of the reaction vessel, where the larger the vessel, the higher the output of the reaction. Time-consuming process development is necessary to fit the synthesis to the various parameters of a larger reactor vessel. A single microreactor can cover a broad range of production scales from mg to kg. For scaling-up, only the operation time of the system is extended and no further process development is necessary.

Microreactors with their small surface-to-volume ratios are able to absorb heat created from a reaction much more efficiently as compared to batch reactor. Since, cooling only takes place at the surface of the reactor, there is a strong temperature gradient from the surface of the reactor to its center. In a microreactor, the heat created by mixing two reagents is also detectable, but the temperature gradient is much smaller. Additionally, it only takes a few millimeters of path length for the reagent stream to cool down again to the temperature outside cooling medium.

One factor that has becoming indeterminate in the use of microreactors is safety. This is dealing with the capability of microreactor in coping with various issues, such as handling the on-site and on-demand production of hazardous or unstable chemicals that can have effects on transport and storage concerns. Efficient heat transfer is also a crucial concern regarding safety. In batch reactors, highly exothermic reactions require extended dosing times. The small inner volume of a microreactor, which is typically less than a milliliter, combined with its strong heat exchange efficiency guarantees the safe and stable performance of highly exothermic reactions over long hours, even for explosive reactants. The extremely exothermic reactions might occur without the same level of cooling commonly found in the conventional apparatus. Most severe cases of reactions that cannot be eagerly performed because they would lead to thermal explosion can be carried out using the microreactor. In microreactor that has little reaction volume in mixtures, heat absorption could be done efficiently, thus guaranteeing its plant security.

In addition, compared to the conventional one, in a microreactor, it is highly possible to control its speed of mixing. This, could consequently enhance the selectivity and yield in the microreaction system [27]. The core part of any microreactor is the mixing regime. Mixing quality is crucial for many reactions where the molar ratio between reactants needs to be controlled precisely in order to suppress side reactions. A sophisticated regime will mix reactants efficiently with a small path length of a few centimeters. Microreactors generally operate in a continuous flow mode. With a reactor volume of less than a milliliter, flow chemistry allows the scale-independent synthesis in bulk amounts in a single day. The small reactor volume facilitates the safe and easy handling of hazardous or unstable materials and highly exothermic reactions. Being an innovative conception in chemical technology, microreactor have facilitated for the introduction of new reaction procedures either in chemistry, pharmaceutical industry, or molecular biology [26].

Reagent streams are continuously pumped into a flow reactor where they are mixed and allowed to react. The product instantly leaves the reactor as a continuous stream. Therefore, the synthesis scale can only be determined by the flow rate and operation time. Identical reactors with an inner volume of less than a milliliter produce kilogram quantities of material when operated for a full day at fast flow rates, or small milligram batches if operated for a few minutes.

Rather than size, other parameters define the performance of a microreactor and will decide if it offers better features than conventional batch reactors. Microreactors are normally composed of metal, glass, or silicon builds. Each material offers specific advantages and disadvantages regarding price and compatibility with reagents or heat conductivity. Glass offers the highest compatibility with aggressive media and reagents. The production of glass microreactors is cheap and offers high visibility. Any blockages or other possible problems in the system can easily be visualized and located through the material.

3.1.2.2 Phase 2: Lab-Scale Novel Ammonia Synthesis System

The first version of the lab-scale novel ammonia synthesis system using magnetic field and induction is shown in Fig. 3.3. It consists of four major subsystems, namely

(a) Reactant gases
(b) Control panel
(c) Magnetic induction zone
(d) Product collecting area

Two gas tanks of pure hydrogen and nitrogen gases were first attached to the system via a ¼ inch stainless steel piping. The pressure of the reactant gases, feed into the system, is controlled by a set of regulators (Wika 216773-1-1) with a pressure range of 0–6 bars. The gas regulator is equipped with a needle valve (TESCOM) to set the flow rate of the feed gases. The volume ratio of reactant gases that flow into the system is controlled using the Sho-Rate flowmeters. The working

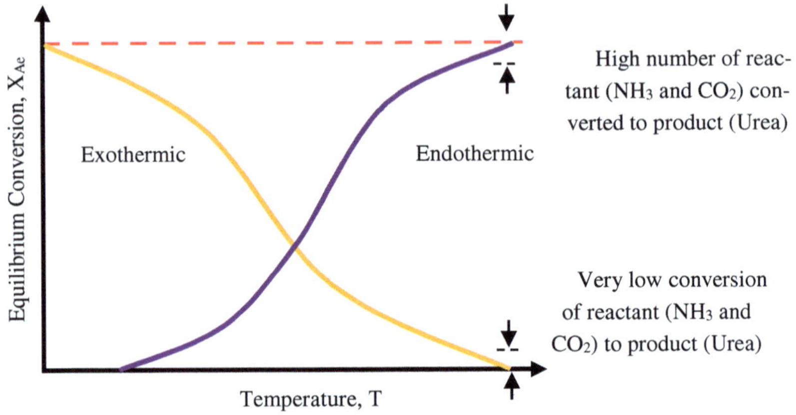

Fig. 3.3 Effect of temperature on equilibrium conversion as predicted by thermodynamics (pressure fixed)

range of the hydrogen and nitrogen flow meter is 0–84 cm^3/min and 0–68 cm^3/min, respectively. This is a standard procedure used.

The innovative approach is when pre-mixed reactant gases pass through an oscillating magnetic field zone, which comprises a series of Helmholtz coils connected to a function generator (Agilent 33250A) to generate an oscillating magnetic field. The main concept is to weaken the gas molecule bond. The oscillating magnetic field is located before the magnetic induction zone (Fig. 3.3) with the intention to increase the yield of ammonia in the magnetic induction zone (main reaction zone) for this system. Several studies were conducted on the effect of ammonia yield with respect to operating temperature were conducted. It is known that magnetization is greatly affected by temperature. The magnetic characteristics of materials can be affected by temperature. An increase in temperature will result in an increase in the magnitude of the thermal vibrations of atoms, which cause the atomic magnetic moments to rotate freely. Therefore, an increase in temperature tends to randomize the directions of any moments that may be aligned and vice versa.

The spin orientation will be distorted by higher temperature especially when the operating temperature is close to Curie temperature. A simple glass tube of 1.0 cm in diameter is used as the microreactor in the system. The glass tube is chosen because glass will not interact with the magnetic field, the applied magnetic field can easily penetrate and pass through the glass to magnetize the nanocatalysts. A customized permanent magnet is employed to create the static magnetic field in the magnetic induction zone. The nanocatalysts are placed in both the oscillating magnetic zone and the magnetic induction zone. The glass tube is placed in the middle of the permanent magnet and at the center of the Helmholtz coils. Glass wool is used to secure the nanocatalysts in the glass tube and to help enhance the mixing of the reactant gases when the gases pass through it. Figure 3.4 shows an overview of the magnetic reaction zone.

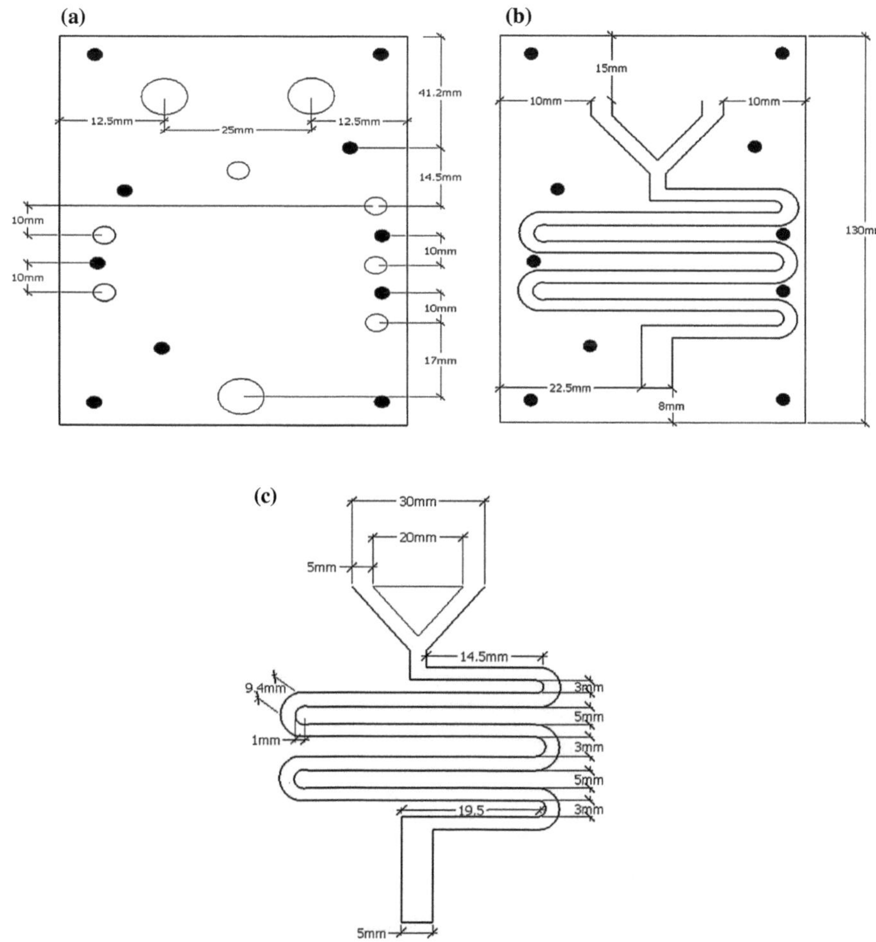

Fig. 3.4 **a**, **b**, and **c** illustrates the miniature reaction systems that can bring additional technical benefit for a variety of objectives. As an illustration, the first microreactor has demonstrated by evidence about its capabilities of a modern concept in which the conventional problems of large-scale synthesis have been removed from consideration at the beginning

After passing through the reacting subsystem, the gases flow into a container with HCl. The ammonia gas flow into the 2.5 mL, 0.01 M HCl solution for 30 min. Ammonia in the form of gas will dissolve in the HCl solution and forms ammonium chloride (NH_4Cl). Quantification of ammonia is performed using the Kjeldahl method and pH calculation method [37].

The permanent magnet strength can be varied from 0T to 0.3T. The oscillating magnetic field is set from 0 Hz to 80 MHz. All the process parameters such as temperature, quantity of Helmholtz coils, frequency of oscillating magnetic field, gas flow rate, gas volume ratio, magnetic induction strength, weight of catalyst used, etc., were set according to the study conditions.

Figure 3.5a shows the ice package used for this study, which surrounds the simple glass tube microreactor and helps lower the testing temperature to 0 C during the experiment. This is to investigate the effects of temperature in this novel method. Figure 3.4b shows a schematic diagram of the ice package tube placed in the magnetic reaction zone setup for the first version of the novel ammonia synthesis system.

Figure 3.6 shows the equipment setup of novel ammonia synthesis system. This design is implemented for the concern of energy efficiency. It is important to recognize that to date, there was little thought to energy requirements in urea synthesis processes. Designing more efficient methods is a necessity and should thus be conducted at room temperature and pressure to reduce energy requirements.

The efficiency of the chemical reactions and the yield of products can be enhanced by the use of nanocatalysts used in this method. Nanocatalytic reagents with great selectivity can be superior to stoichiometric reagents. New catalysts and more emphasis on nanocatalytic processes are the future of green urea synthesis system.

Fig. 3.5 First version of the lab-scale novel ammonia synthesis system

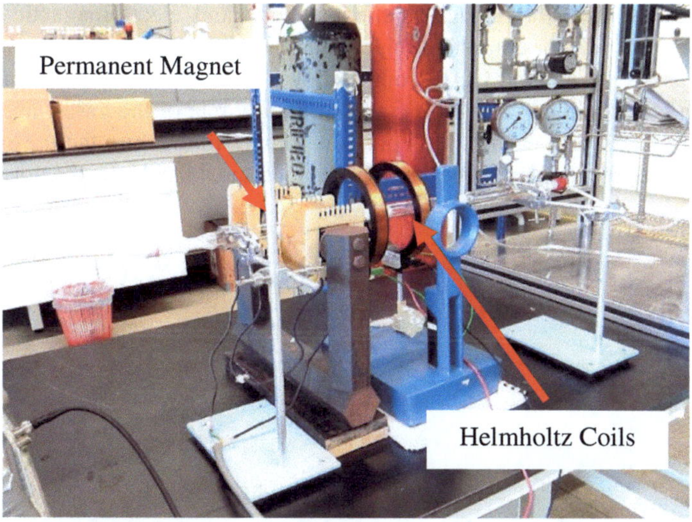

Fig. 3.6 Magnetic reaction zone of the novel ammonia synthesis system

3.1.2.3 Phase 3: Pilot-Scale Novel Urea Synthesis System

The success of the first reactor in producing ammonia at low temperature and ambient pressure enables the lab-scale synthesis of urea in presence of static magnetic and electromagnetic field. This has been carried out using the reactor presented in Figs. 3.7 and 3.8. The reactor consists of two main parts; The first is

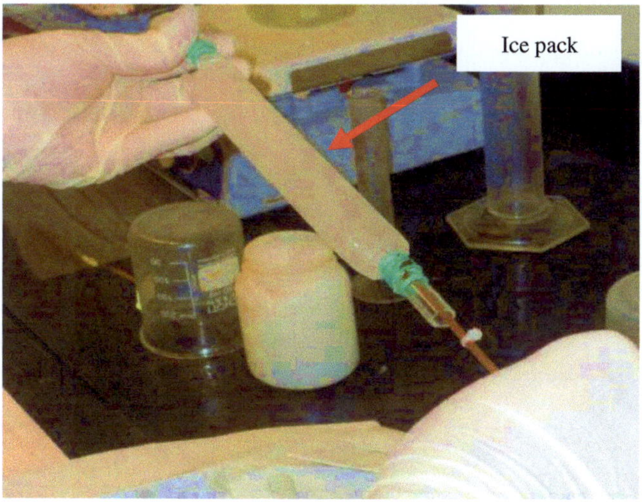

Fig. 3.7 **a** Loading of the nanocatalyst into the reactor surrounded by the ice pack

the gas reactor, and the second is the magnetic and electromagnetic induction sources. The gas reactor consists of nine main parts as follows: gas cylinders for the reactant gases (H_2, N_2, and CO_2), control valves to control the gases that will pass through the reactor, flow meters to control the flow rate of the gases, mixer to mix the reactant gases with each other's pressure gauges to measure the pressure inside the reactor, thermocouples to measure the temperatures inside the reactor, tube pipe

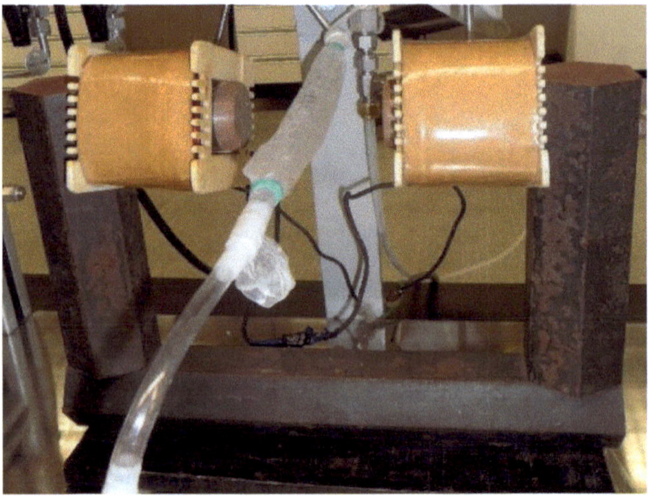

Fig. 3.8 **b** Ammonia synthesis at 0 °C

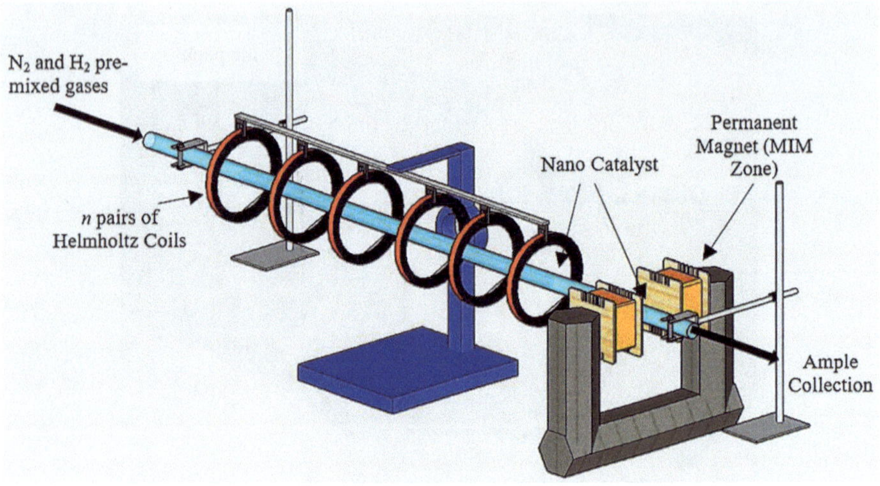

Fig. 3.9 Schematic diagram of the equipment setup for the first version of novel ammonia synthesis system

which is a space where the reaction takes place, catalyst holder where the nanocatalyst is placed, back pressure controller to control the pressure inside the reactor, and vessel to collect and dilute the urea samples (Fig. 3.9).

The magnetic induction sources consist of static magnetic field source and electromagnetic emitter. The static magnetic field surrounds the nanocatalyst for enhancing the catalytic activity, while the electromagnetic field weakens the bonds between the molecules of urea reactant gases.

No.	Parts name
1	Gas cylinder
2	Control valve
3	Flow meter
4	Mixer
5	Pressure gauge
6	Thermometer
7	Tube pipe
8	Electromagnetic emitter
9	Nanocatalyst holder
10	DC magnetic field
11	Back pressure controller
12	Collecting vessel

A mixture of pure hydrogen, nitrogen, and carbon dioxide gases pass over the nanocatalyst which is placed between two poles of static magnetic fields. The reactant gases were exposed to EM field with different frequency values. The chemical reaction occurs at ambient temperature and pressure where the strength of applied magnetic field is fixed to 1 T. The flow rates of H_2, N_2, and CO_2 were adjusted at different ratios (3:1:1, 3:1:2, and 3:1:3). For every run, urea sample was collected after 45 min by diluting it in acetonitrile (ACN). The FTIR device was used to quantify and qualify the produced urea. The experiment parameters and their values are shown in Table 3.1.

Table 3.1 Experiment parameters for urea synthesis using EM field at lab-scale

No	Parameter	Value
1	Magnetic field (T)	1.0
2	Electromagnetic wave frequency (GHz)	0.25, 0.5, 0.75, 1.0, 1.25, 1.5, 1.75, 2.0, 2.25, 2.5, 2.75, 3.0, 3.25,3.5, 3.75, 4.0, 4.25, 4.5, 4.75 and 5.0
3	H_2 flow rate (L/min)	0.3
4	N_2 flow rate (L/min)	0.1
5	CO_2 flow rate (L/min)	0.1, 0.2 and 0.3
6	Nanocatalyst	α-Fe_2O_3

3.1.2.4 Phase 4: Pilot-Scale Synthesis of Urea

Green urea has also been synthesized using a pilot-scale reactor. The schematic diagram and the experimental setup of the reactor are shown in Figs. 3.10 and 3.11, respectively. The reactor consists of five main parts, namely: the catalyst reactor with the microwave generator, the distillation column, the vacuum evaporator column, the fluidized bed dryer reactor, and the storage column (Fig. 3.12 and 3.13).

The ammonia and carbon dioxide gases flow inside the reactor at a rate of 1L/min each. The mixture of ammonia and carbon dioxide is pumped from the gas booster which maintains the reaction pressure in the catalyst reactor. The mixture is preheated to 200 °C before passing over 0.5 kg of hematite nanocatalyst placed inside a chamber that is connected to the microwave generator. The chemical reaction takes place at a temperature of 200 °C and a pressure of 50 bar, where the frequency of applied electromagnetic wave is fixed to 3.25 GHz. The product (molten urea, ammonia, and carbon dioxide) then goes to the distillation column to separate unreacted gases to be exhausted or recycled. The molten urea that settles down at the bottom of the column is sent to a vacuum evaporator column for drying and moisture removal. After that, the product is transferred to the fluidized bed, where the drying and coating of urea is conducted. Finally, the coated urea is sent to the storage column for storage purposes. The urea product collected is then quantified and qualified using FTIR. The experiment parameters and their values are shown in Table 3.2.

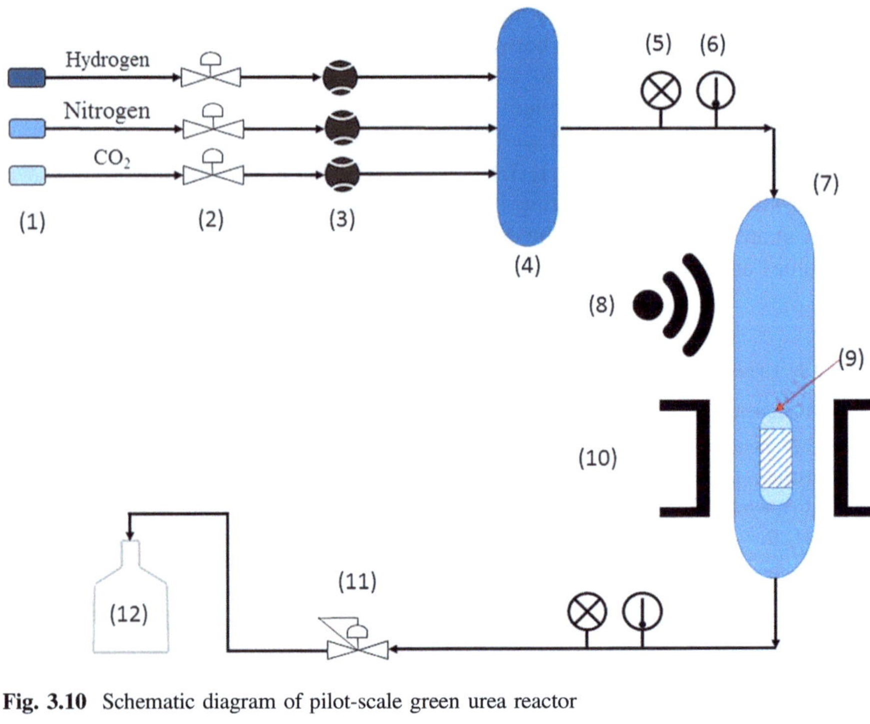

Fig. 3.10 Schematic diagram of pilot-scale green urea reactor

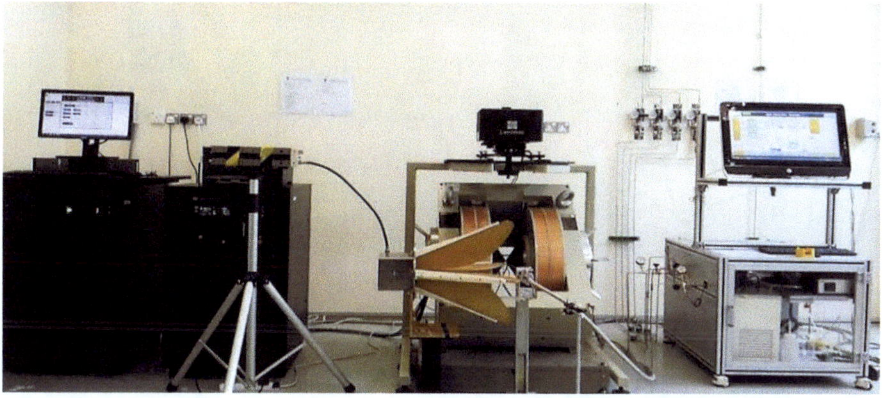

Fig. 3.11 Experimental setup of lab-scale urea reactor with static magnetic field source and electromagnetic emitter

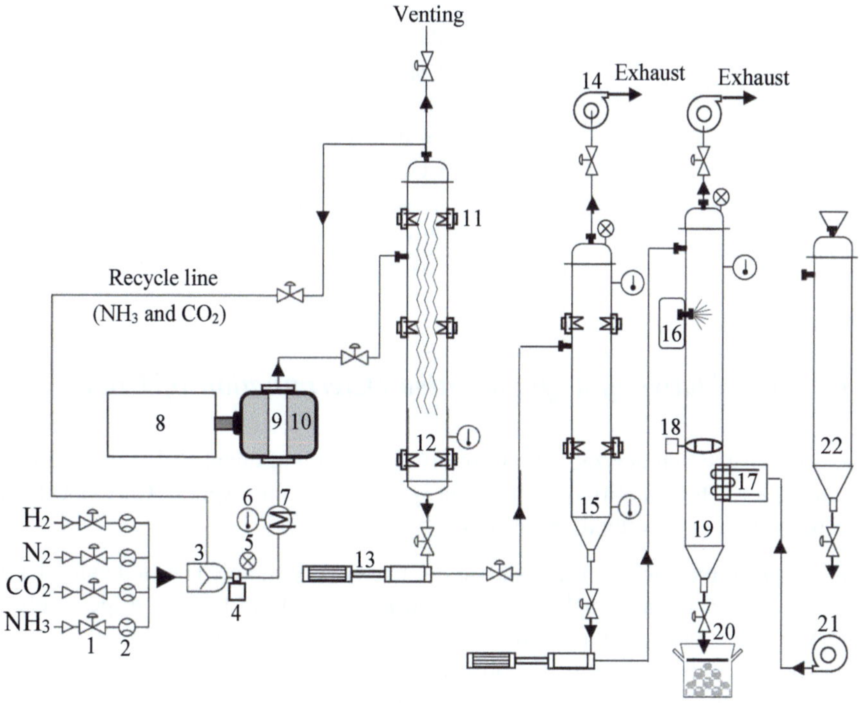

Fig. 3.12 Schematic diagram of pilot-scale green urea reactor

Fig. 3.13 Experimental setup of pilot-scale green urea reactor

Table 3.2 Experiment parameters for urea synthesis using EMF at pilot-scale

No	Parameter	Value
1	EMF frequency (GHz)	3.25
2	H_2 flow rate (L/min)	3
3	N_2 flow rate (L/min)	1
4	CO_2 flow rate (L/min)	1
5	Temperature (°C)	200
6	Pressure (bar)	50
7	Nanocatalyst	α-Fe_2O_3

3.1.3 Qualitative and Quantitative Determination of Urea

The yield of the synthesized green urea is qualified and quantified using the FTIR spectrometer. The quantity of green urea is assessed using Agilent Cary 630 FTIR. The procedure is described as follows:

i. 50 μL of the solution (urea and ACN) from every sample is tested using FTIR.
ii. The row data and graphs are exported in asp and pdf format, respectively.
iii. Various concentration of urea in acetonitrile are prepared.
iv. The calibration curve of the absorbance and the concentration is performed.
v. Finally, the yield of urea is calculated by comparing the values of urea peaks with the peaks of the calibration curve

3.1.3.1 Experimental Design

Response Surface Methodology (RSM) from the design expert software was used to design the research experiment. The design was developed to investigate the relationships between the input variables, which is electromagnetic field frequency with CO_2 flow rate and the response of the urea yield. This design aims to determine the optimum variable values that contribute toward obtaining the highest (optimum) response. The design of the experiments consist of four major parts:

3.1.3.2 Building the Design

Building the design of the experiment initiates by choosing the design method and type to fit the experiment parameters. This is followed by the experiments conducted to fill the response results. Lastly, choosing the suitable model terms was conducted to offer a logic demonstration of the experiment results.

RSM with optimal (custom) design was chosen to generate the experimental design for urea synthesis. The numerous factors refer to the number of variables, which are two variables (A) electromagnetic field frequency in Hertz and (B) flow rate of CO_2 in liters per minute. The levels of the variables were chosen to be 20 for A and 3 for B. The range of A and B variables input was set to 0.25–5 GHz and 0.1–0.3 L/min, respectively.

3.1.3.3 Choosing the Model Terms

The software provides various options that can be used for the process order. In this design, a modified fifth process order was selected for the module. The modification excludes all p-values that were higher than 0.1. It is necessary that the response becomes well-fit with the selected process order.

3.1.3.4 Evaluating the Design

The design was evaluated using the data analysing and diagnostic tools in the software such as analysis of variance (ANOVA) tables and diagnostic plots. Both tools aid to estimate the validity of design.

3.1.3.5 Optimization

The optimization can be carried out using various methods, such as numerical or graphical. In this design, numerical optimization was chosen. The optimization was performed for the variables and the response by choosing the goals for each one. The maximum and minimum responses were chosen as the range for the variables.

Next, a list of the optimum possible solutions was generated from the software to obtain the optimum output (response) for the design at hand.

3.1.4 Numerical Analysis of Urea

3.1.4.1 Qualitative and Quantitative Analysis of Urea

An FTIR spectrometer was employed to qualify and quantify the green urea. The yield was calculated by comparing the intensity of absorbance peaks for the synthesized green urea with commercial pure urea. Figure 3.5 demonstrates the calibration curve for the commercial pure urea, which illustrates the relationship between the absorbance and the concentration of pure urea at various values, namely, 1×10^3, 2×10^3, 3×10^3, 4×10^3, and 5×10^3 ppm.

Figure 3.14 shows the FTIR absorbance spectra for two samples (green urea samples and pure acetonitrile). According to the figure, it is clear that there is a difference in the intensity for the samples in the ranges (3300–3700 cm^{-1}), (1600–1700 cm^{-1}), and (700–1200 cm^{-1}), which agrees with infrared absorption bands of urea with frequencies (3333–3436), (1659–1676), 1630, 1595, (1464–1492), (1000–1014), and 793 (cm^{-1}) corresponding to NH (stretching), CO (stretching), NH_2, NH (bending), CN (bending), CN (stretching), and OCNN groups, respectively [33, 34].

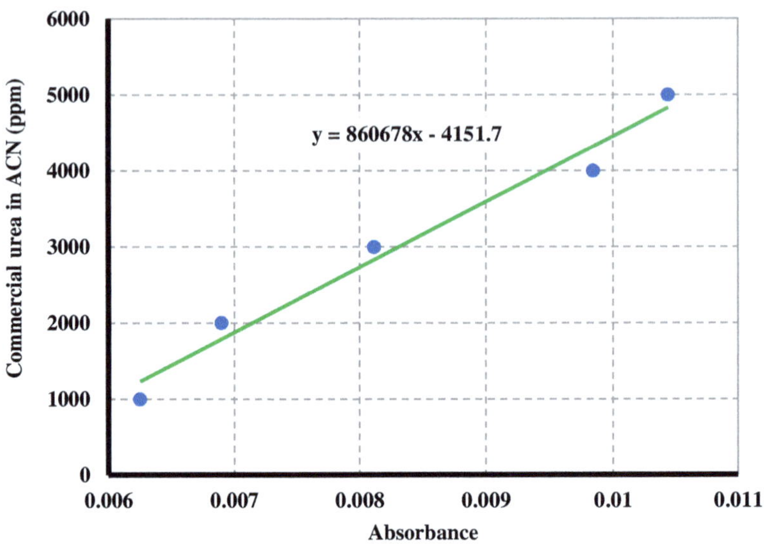

Fig. 3.14 Concentration calibration curve of pure commercial urea

3.1.5 *Response Surface Methodology (RSM) Results*

Response Surface Methodology (RSM) was employed as the analysis method to study the relationship between the variables and the response. Two variables were investigated, namely, (A) electromagnetic field frequency and (B) flow rate of CO_2 gas where the response was the green urea yield.

3.1.5.1 Model Fitting and Analysis of Variance

Analysis of Variance (ANOVA) is a hypothesis testing procedure used to determine whether the variables significantly affect the response values. ANOVA is also used to evaluate the research design. F-values were calculated to measure the size of the effects by comparing the ratio of the differences between the variables mean.

The impact of the variable depends on the F-value. When the value of F is large this means that the variable has a real and significant effect on the value of the response in the model. If the F-value ratio in the model gives no effect this means that the model has no effect and the hypothesis is thus rejected. Table 3.3 shows the statistics that are used to test the hypotheses about the variables mean of the design.

The F-value of the model is 101.0 implying that the model is significant. There is only a 0.01% chance that an F-value could occur due to noise. The probability (p-value) for A, B, A^3, and A^5 terms is less than 0.05 indicating that these model terms are significant.

The R-squared statistic value indicates that the model as fitted can explain 96.34% of the variability in the urea yield. This implies that 96.34% of the experiment values for urea synthesis are reliable. The adequate precision ratio is 30.62 implying that the model has high signal-to-noise ratio which qualifies it to navigate the design space.

Table 3.3 ANOVA results for the experimental design of urea synthesis using electromagnetic field

Source	Sum of squares	df	Mean square	F-value	p-value Prob > F
Model	502.2×10^6	6	83.7×10^6	101.0	<0.0001
A-EM frequency	9.0×10^6	1	9.0×10^6	10.9	0.0031
B-CO_2 flow rate	5.4×10^6	1	5.4×10^6	6.5	<0.0177
A^2	4.0×10^6	1	4.0×10^6	4.8	0.0384
B^2	379.1×10^6	1	379.1×10^6	457.7	<0.0001
A^3	11.2×10^6	1	8.9×10^6	8.0	0.0102
A^5	13.2×10^6	1	13.2×10^6	13.5	0.0012
Residual	19.0×10^6	23	0.8×10^6		

The model used for urea yield prediction is described by Eq. 3.4 in terms of actual factors.

$$\text{Urea yield} = -27025.9 - 2594.7f^2 + 412409.0\,v + 2334.4f^2 - 1013520.0\,v^2 - 552.9f^3 + 7.8f^5 \tag{3.14}$$

where f is the electromagnetic wave frequency in Hz, v is the CO_2 flow rate in L/min.

It is crucial to know that Eq. 3.4 in terms of actual factors can be used to make predictions about the response (urea yield) for given levels of each factor. Here, the levels should be specified in the original units for each factor.

3.1.5.2 Diagnostic Model

The diagnostic plots were implemented to evaluate the two designs and ensure that the statistical assumptions tally with the analysis data. Figure 3.7 shows the normality plot of the residuals for the design. Most residuals lie on the straight line which means that the distribution of residuals is normal. The p-value on the Anderson–Darling test is greater than 0.005 which also confirming that the distribution is normal.

Figure 3.15 depicts a plot of predicted against actual values for the response (urea yield) values in the design. In this plot, the straight line indicates the predicted values while the small boxes represent the actual experimental values of the urea yield. The plot reflects that the actual values are located near the straight line and have a correlation coefficient R^2 of 0.9634, confirming the accuracy of the model (Fig. 3.16).

Residuals versus predicted response for urea yield are depicted in Fig. 3.15. The graph shows that almost all points lie within area ± 3.0. This indicates that the assumption of constant variance is confirmed, suggesting that the model is suitable.

3.1.6 Urea Yield Analysis

The relationship between the two-independent variables (EM field frequency and CO_2 flow rate) with the response (urea yield) was studied with use of the software. In every case, only one variable from the two variables was studied while the second remained fixed.

Figure 3.17a, b, and c, shows the relationship between the electromagnetic wave fixed at 0.1, 0.2 and 0.3 L/min, respectively.

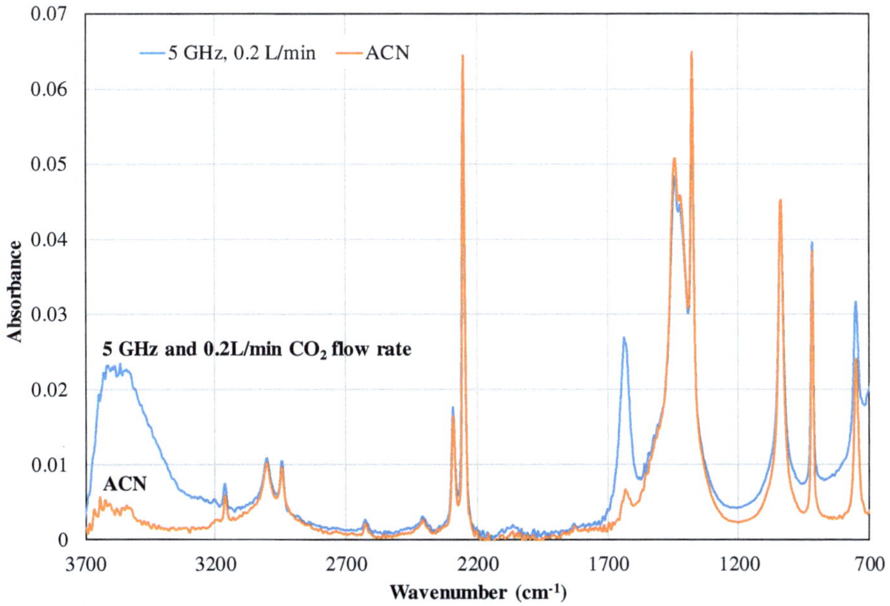

Fig. 3.15 FTIR absorbance spectra of pure ACN and urea sample synthesized by using EM field with 5 GHz frequency and 0.2 L/min CO_2 flow rate

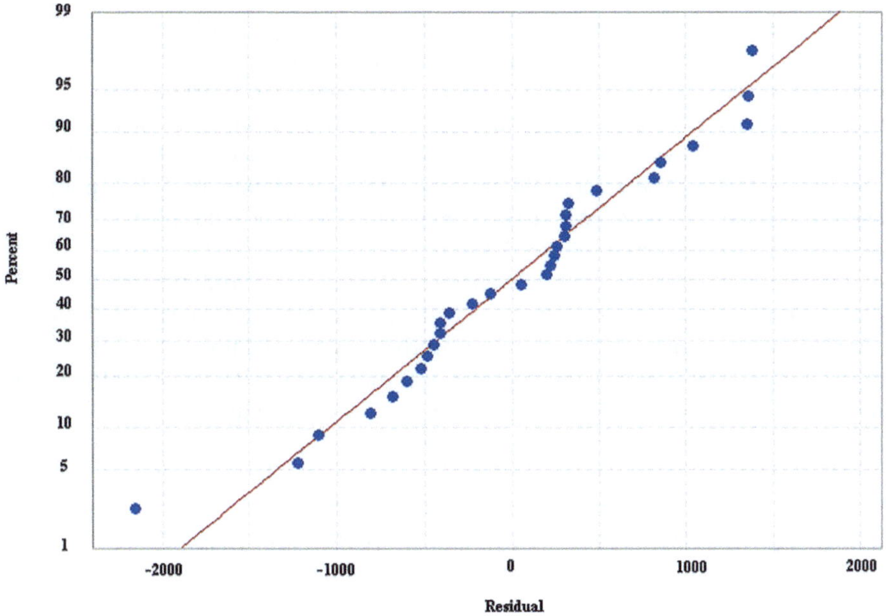

Fig. 3.16 Normality plot of the residuals for the design

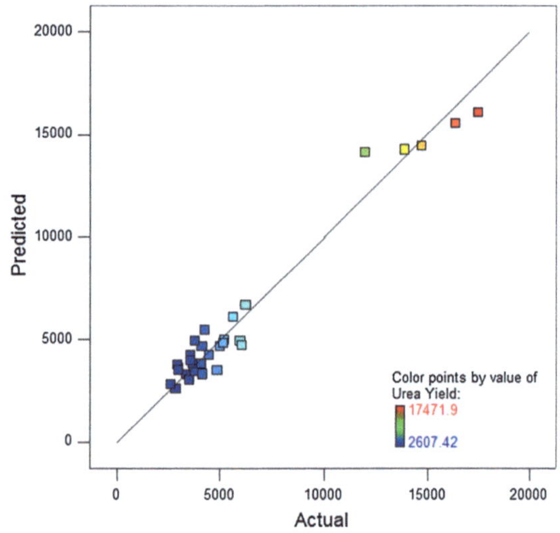

Fig. 3.17 Plot of predicted versus actual values for the design

According to the graphs, the urea yield was affected by the electromagnetic wave frequency and the carbon dioxide flow rate. Figure 3.10a shows that the urea yield increased with an increase in the electromagnetic wave frequency until it reached a peak at 1 GHz, and then fell to a minimum of 2 GHz before rising to another peak at 3 GHz. It fell and rose again until it reached a maximum peak of about 6000 ppm at 5 GHz.

The same occurs when the flow rate of CO_2 increased to 0.3 L/min, but at a CO_2 flow rate of 0.2 L/min, a higher maximum yield of about 16000 ppm at 5 GHz was obtained as shown in Fig. 3.10b. On the other hand, Fig. 3.10d shows the relationship between the carbon dioxide gas flow rate and the urea yield when the electromagnetic wave frequency and H_2 flow rate were fixed at 5 GHz and 0.3 L/min, respectively. The urea yield increased to a maximum of 16000 ppm at a carbon dioxide gas flow rate of 0.2 L/min and was then reduced with a further increase in CO_2 flow rate up to 0.3 L/min.

Figure 3.17a, b shows three-dimensional surface and two-dimensional contour plots, respectively, for the interaction between the response (urea yield) and the variable (electromagnetic frequency) respectively, when the flow rate of H_2 is fixed to 0.3 L/min. The figures show that the peak urea yields of over 15000 ppm (red region) are achieved when the electromagnetic waves frequencies were 1, 3, and 5 GHz, with a CO_2 flow rate of 0.2 L/min (Fig. 3.18).

This high yield at the particular three frequencies may be due to the electron spin resonance and interconversion of spin multiplets through the action of inhomogeneous magnetic fields, hyperfine interactions, Stark effect [26, 30], which can be explained as follows. The electromagnetic field affected the gases by electric and magnetic fields. The magnetic field also affects electron spins of the three gases.

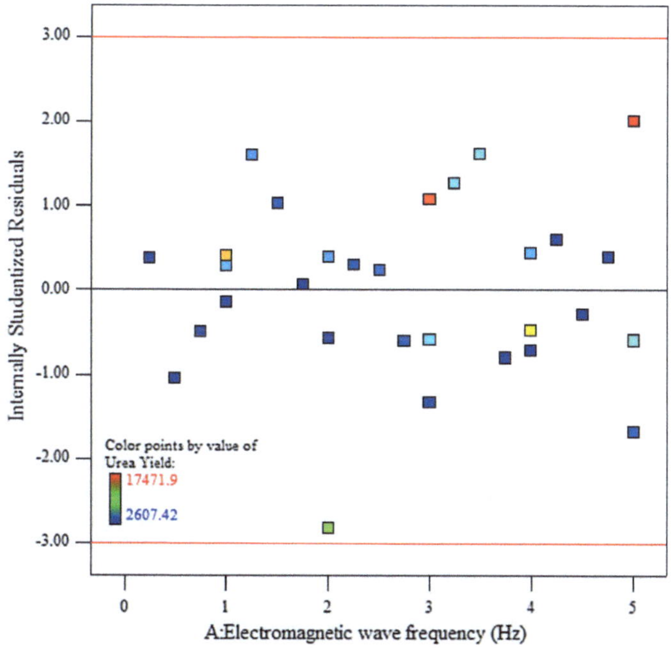

Fig. 3.18 Plot of residuals versus electromagnetic field frequency for the design

The flipping of electron spins weakens the bonds between gas molecules which leads singlet to triplet to occur [20], while the electric field enhances the chemical reactions probability, modifies reactive scattering resonances and causes shifting in the energy levels of atoms (Stark effect) [29].

For a singlet to be converted to a triplet in the catalyst, inhomogeneous magnetic fields must be applied [28]. In the current research, the catalysts were exposed to homogeneous applied magnetic field, but this could be considered inhomogeneous because of the density distribution difference, orientation, geometric factor, and the morphology of nanowires. Sharp tips of nanowires occupy high electron density which leads to better induced magnetic fields [5]. The longer the distance between the nanowires, the higher the magnetic field gradient, and vice versa. Randomly oriented nanowires lead to different direction and strengths of induced magnetic field which will affect the net magnetic field strength [35]. During the interaction between gases and nanocatalysts, the wave function of nanocatalyst tips overlaps with the wave function of the hydrogen, nitrogen, and carbon dioxide gases, which weakens or breaks the bonds in the gases due to interference [38].

The strength of static applied magnetic field is important to prevent a singlet state from mixing with the triplet states [28]. In a strong magnetic field where the frequency is much higher than hyperfine coupling constant, the singlet to triplet transition can occur while the secular approximation of the hyperfine coupling is also applicable. However, in a weak magnetic field where the frequency is much lower than hyperfine coupling constant, the non-secular terms of the hyperfine couplings become significant and the singlet states are mixed with all triplet states [28].

3.1.7 Optimization of Green Urea

The input and output values were constrained by determining the goals as follows. In term of the design, the two variables A (electromagnetic field frequency) and B (CO_2 flow rate) were chosen to be in range option and for the response (urea yield) to be maximum. The urea yield was set at a maximum in order to achieve the highest yield of urea.

Table 3.4 shows the determined constraints that the software was followed during run the optimization.

A list of the optimum possible solutions was generated by the software to obtain the optimum output (response). Table 3.5 depicts some of the optimum conditions that were suggested by the software to achieve the optimal yield of urea.

Table 3.4 Constraints of the optimization for the design

Name	Unit	Goal	Lower limit	Upper limit	Lower weight	Upper weight
A: Electromagnetic field frequency	Hz	in range	50	5×10^9	1	1
C: CO_2 flow rate	L/min	in range	0.1	0.3	1	1
Urea yield	ppm	Maximize	0	18000	1	1

Table 3.5 Experimental and predicted yield of urea for the design

Test no	EM field frequency (GHz)	CO_2 flow rate (L/min)	Predicted urea yield (ppm)	Experimental urea yield (ppm)	Experimental actual urea yield (ppm)	Percentage error (%)	Desirability (%)
1	5.00	0.203	16098	15032	16501	2.5	89
				17000			
				17472			
2	3.163	0.203	16173	15032	15489	4.2	87
				16301			
				15134			

3.1.8 Pilot-Scale Urea Product

FTIR spectrometer was used to qualify the green urea. Figure 3.19 shows the FTIR absorbance spectra of commercial pure urea and the synthesized green urea. According to the figure, there is a high match between the peaks of green urea and commercial urea which confirms that the green urea was successfully synthesized. Moreover, the green urea has higher intensity in the absorbance spectra than the commercial urea (Fig. 3.20).

The green urea has sharp peaks in the wavenumbers 717, 793, (1000–1014), (1050–1060), (1150–1160), (1464–1492), 1595, (1659–1676), (3250–3268), and (3333–3436) (cm^{-1}) which correspond to NH, OCNN, CN (stretching), NH_2

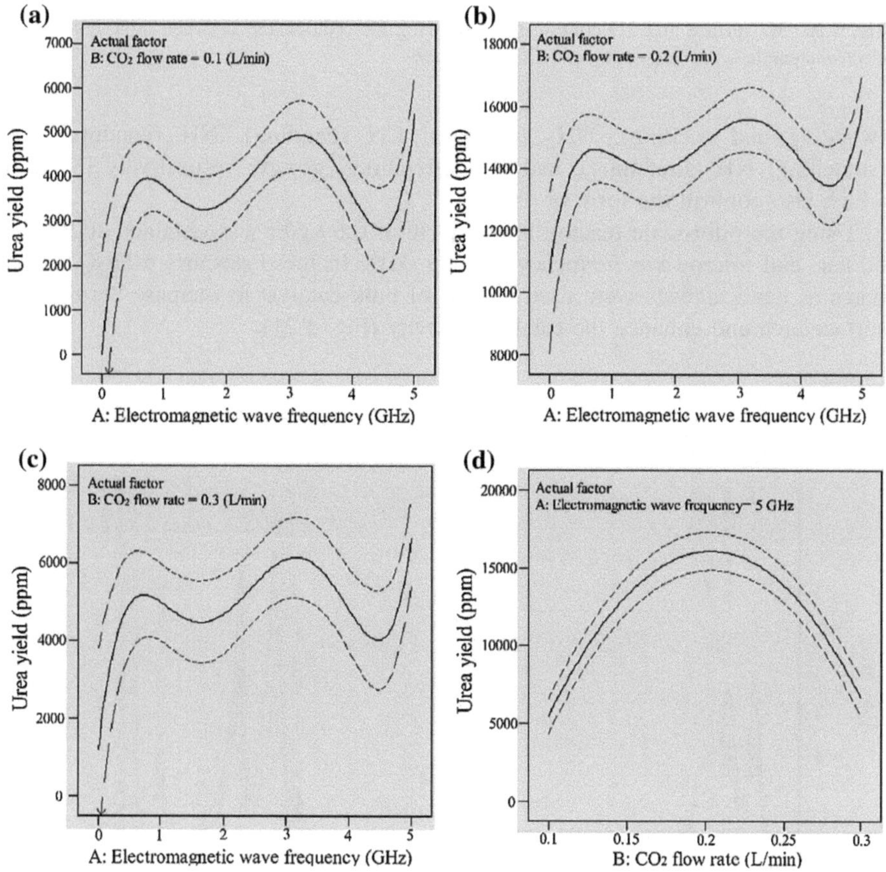

Fig. 3.19 **a**, **b**, and **c** Urea yield against electromagnetic wave frequency at 0.1, 0.2, and 0.3 L/min CO_2 flow rate, respectively. **d** Urea yield against CO_2 flow rate

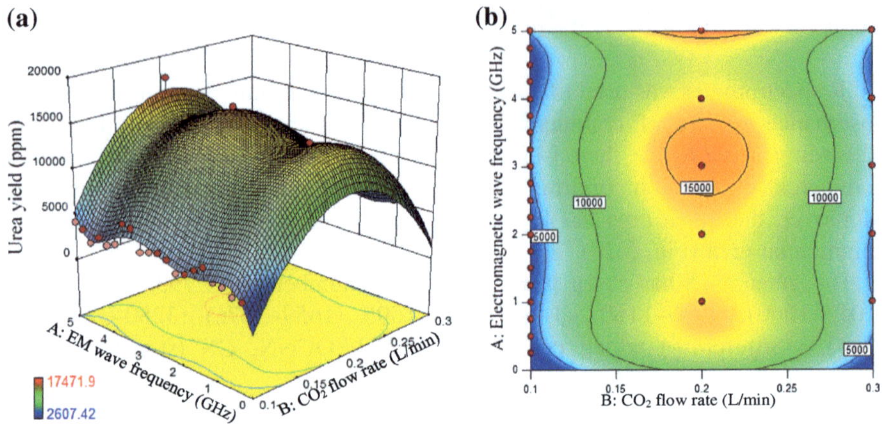

Fig. 3.20 3D surface and 2D contour plots showing the relationship between urea yield versus electromagnetic wave frequency and CO_2 flow rate

(wagging and rocking), NH_2 (rocking), CN (bending), NH (bending), CO (stretching) NH (stretching), and NH (stretching) groups, respectively [33, 34], which also confirm the form of urea.

Using the pilot-scale reactor, a urea yield of 0.6 kg/hr was obtained at 200 °C, 50 bar, and microwave frequency of 3.25 GHz. In these reactors α-Fe_2O_3 nanowires as nanocatalysts were used instead of bulk catalyst to increase the reaction surface area and enhance the catalytic activity (Fig. 3.21).

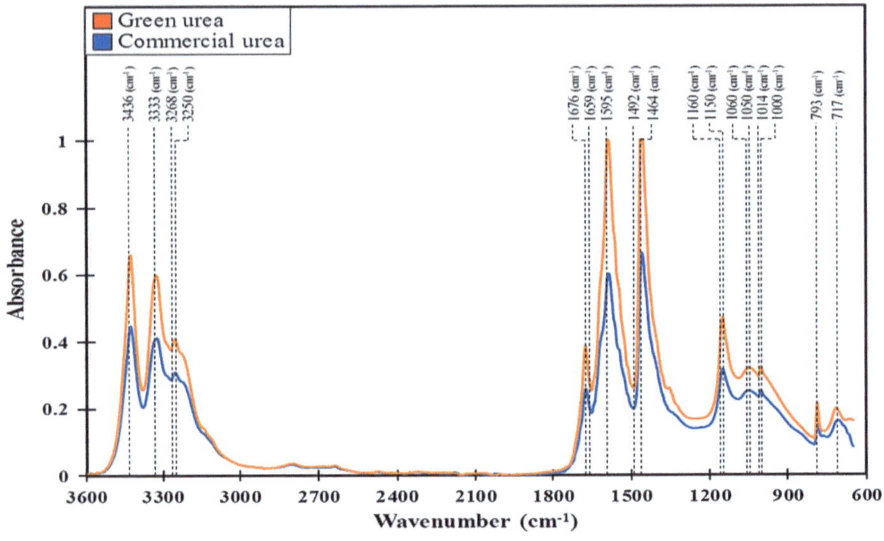

Fig. 3.21 FTIR absorbance spectra of commercial and green urea

The microwave consists of both electric and magnetic fields, where the oscillating magnetic field enhances the singlet to triplet conversion [20], while the electric field enhances the chemical reactions probability, modifies reactive scattering resonances and causes shifting in the energy levels of atoms (Stark effect) [29].

3.1.9 The Patent of Urea Synthesis Process

The patent summary of urea synthesis is shown in Table 3.1. As seen in Table 3.6, the method employed to produce urea involves the use of high-temperature heat combustion and gas fired burner boilers. This process contributes to the high emissions of NO in air at high temperature in the burner which then adds more pollution to the surrounding atmosphere. The combusted NO is generated as a function of air to fuel ratio and is more pronounced when the mixture is on the fuel-lean side of the stoichiometric ratio, which is determined by the ratio of chemicals in the reaction. However, none of the methods patented for urea synthesis [39–49] used the green technology of urea production such as the electromagnetic induction and nanoparticle material.

3.1.10 Summary

Microreactor design was implemented for the concern of energy efficiency. This method is a necessity and should be conducted at room temperature and pressure to reduce energy requirements. The overall efficiency of the chemical reactions and the yield of products can be enhanced by the use of nanocatalysts. The emphasis on nanocatalytic processes is the future of green urea synthesis system. The microreactor technology provides benefits to future sustainability such as scale-independent synthesis, product profile improvement, accelerated process development, enhanced safety, constant product output quality, cleaner product profile, and higher yields. Based on the patent summary of urea synthesis, the high-temperature heat combustion and gas-fired burner boilers methods are employed for the production of urea. These processes contribute to the high emissions of NO in air and thus add more pollution to the surrounding atmosphere. However, none of the methods patented for urea synthesis implemented the green technology approach such as the electromagnetic induction and nanoparticle material, as proposed by this green urea study. Interdisciplinary research involving chemists, physicists, engineers, material scientists, and bioscience technologists has expanded the fields of green urea technology to enhance our future sustainability.

Table 3.6 Patent summary of urea synthesis process

Urea synthesis process: the patent summary update				
Ref.	Patent name and no.	Year published	Description of patent	Patent no and details
[39]	UREA SYNTHESIS PROCESS Ivo Mavrovic US3172911 A	Mar 9, 1965	This invention relates to the synthesis of urea from ammonia and carbon dioxide. A process modification has been developed, whereby unconverted reactants may be recycled to the synthesis step while avoiding the recycle of excess water	AGENT United States Patent 3,172,911 UREA SYNTHESIS PROCESS Ivo Mavrovic, New York, N.Y., assignor to Chemical Construction Corporation, New York, N.Y., a corporation of Delaware Filed June 16, 1961, Ser. No. 117,631 Claims. (Cl. 260-555)
[40]	L. H COOK UREA SYNTHESIS PROCESS US3816528 A	Jun 11, 1974	A urea synthesis process employing ammonia and carbon dioxide as feed reactants in which mixed off-gas derived from ammonium carbamate decomposition is compressed together with feed carbon dioxide to urea synthesis pressure and reacted exothermically with feed ammonia while in heat exchange with urea synthesis stream, which is concomitantly stripped with intermediate pressure recycle mixed gas	AGENT United States Patent O 3,816,528 UREA SYNTHESIS PROCESS Lucien H. Cook, Port Washington, N.Y., assignor to Chemical Construction Corporation, New York, N.Y. Filed Nov. 16, 1970, Ser. No. 89,754 Int. Cl. C07c 127/00 U.S. Cl. 260-555 A 6 Claims
[41]	UREA SYNTHESIS PROCESS US3200148 A	Aug 10, 1965	This invention relates to the synthesis of urea ammonia and carbon dioxide. An improved method for total recycle of unconverted reactants has been developed, which accomplishes direct recycle by adiabatic compression of the unconverted reactants as a hot gas stream, and thereby permits the recovery of process reaction heat as usable process steam	UREA SYNTHESIS PROCESS Filed Aug. 29, 1962 LUCIEN H. COOK IVO MAVROVIC INVENTORS AGENT United States Patent 3,260,148 UREA SYNTHESIS PROCESS Lucien H. Cool-1, Port Washington, and Ivo Mavrovic, New York, N.Y., assignors to Chemical Construction Corporation, New York, N.Y., a corporation of Delaware Filed Aug. 29, 1962, Ser. No. 220,339 l Claims. (Ql. 260-4555)

(continued)

Table 3.6 (continued)

Urea synthesis process: the patent summary update				
Ref.	Patent name and no.	Year published	Description of patent	Patent no and details
[42]	UREA SYNTHESIS PROCESS US3668250 A	Jun 6, 1972	In the present invention, the off-gas generated from the urea synthesis effluent stream by heating at reduced pressure is mixed for reabsorption and reabsorbed in an aqueous absorbent solution prior to recycle in a combination recycle urea synthesis process, by a process which efficiently uses the heat generated by off-gas absorption to heat the product aqueous urea solution and thereby evaporate water, by indirect heat exchange of the respective streams	Inventors: Karafian Maxim Original Assignee: Chemical Construction Corp
[43]	J. H. MARTEN UREA SYNTHESIS PROCESS US3354205 A	Nov 21, 1967	The present invention relates to the synthesis of urea, by the reaction of ammonia and carbon dioxide at elevated pressure. An improved complete recycle procedure is provided, which features an improved method for the economical removal of inert from the system. It has been determined that essentially complete recovery of ammonia vapor from the vented inert may be successfully accomplished by scrubbing the inert with pressurized aqueous ammonium carbamate solution, derived from a later stage of the complete recycle process	AGENT United states Pater 3,354,205 Patented Nov. 2 l, 1967 ice 3,354,205 UREA SYNTHESIS PROCESS Jerome H. Marten, Nixon, NJ., assigner to Chemical Construction Corporation, New York, NX., a corporation of Delaware Filed Mar. 26, 1964, Ser. No. 354,983 2 Claims. (Cl. 2150–555)

(continued)

Table 3.6 (continued)

Urea synthesis process: the patent summary update				
Ref.	Patent name and no.	Year published	Description of patent	Patent no and details
[44]	Urea synthesis process US4296252 A	Oct 20, 1981	An improvement in the cyclic process for producing urea wherein CO_2 and NH_3 are reacted in the presence of an aqueous ammoniacal solution in a urea synthesis reactor at an elevated temperature and at an elevated pressure in excess of 1800 PSIG to form a urea synthesis reactor effluent fluid at high pressure	Inventors: Ivo Mavrovic Original Assignee: Ivo Mavrovic
[45]	L. H. COOK UREA SYNTHESIS PROCESS US 3370090 A	Feb 20, 1968	The present invention relates to the synthesis of urea from ammonia and carbon dioxide at elevated temperature and pressure, in which the by-product off-gas consisting of unconverted process components is completely recycled to the synthesis step in the gaseous state by means of adiabatic compression. An improved process sequence was devised, in which the gaseous carbon dioxide feed stream is added to the off-gas stream prior to compression, thus providing improved efficiency in adiabatic compression	AGENT Unite rates Patent Ofi Patented Feb. 20, 1968 ice 3,370,090 UREA SYNTHESIS PROCESS Lucien H. Cook, Port Washington, N.Y., assignor to Chemical Construction Corporation, New York, N.Y., a corporation of Delaware Filed Sept. 30, 1965, Ser. No. 491,675 16 Claims. (Cl. 260–555)
[46]	UREA SYNTHESIS PROCESS IVO MAVROVIC IN VENTOR US 3527799 A	Sep 8, 1970	A urea synthesis process improvement is provided, in which the pressure of the synthesis autoclave effluent stream is adiabatically reduced to produce an off-gas rich in free ammonia, which is separated from the residual liquid stream containing product urea	AGENT United States Patent 3,527,799 UREA SYNTHESIS PROCESS Ivo Mavrovic, New York, N.Y., assignor to Chemical Construction Corporation, New York, N.Y., a corporation of Delaware Filed Jan. 20, 1966, Ser. No. 521,921 Int. Cl. C07c 127/00 U.S. Cl. 260–555 3 Claims

(continued)

Table 3.6 (continued)

Urea synthesis process: the patent summary update				
Ref.	Patent name and no.	Year published	Description of patent	Patent no and details
[47]	Urea synthesis process US 6518457 B1	Feb 11, 2003	A process for the synthesis of urea, which comprises subjecting a urea synthesis solution from a urea synthesis column to stripping with a raw material carbon dioxide under a pressure approximately equal to the urea synthesis pressure to separate a major portion of unreacted ammonia and carbon dioxide as a mixed gas from the urea synthesis solution, condensing the mixed gas by indirect heat exchange with the urea synthesis solution from which the mixed gas was separated and that is preferably decompressed to a pressure lower than the urea synthesis pressure, and heating the urea synthesis solution by heat of condensation generated at the time of the condensation of the mixed gas	Inventors: Eiji Sakata, Yasuhiko Kojima Original Assignee: Tokyo Engineering Corporation
[48]	Urea synthesis process US 3579636 A	May 18, 1971	Ammonia and carbon dioxide are reacted in a reactor at elevated pressure and elevated temperature to form urea. In this synthesis, an intermediate compound, ammonium carbamate, is formed which then loses one molecule of water to yield urea. Only a part of the carbon dioxide charge is converted to urea, and the remainder is present in the reactor effluent as unconverted ammonium carbamate	UREA SYNTHESIS PROCESS Filed May 16, 1968 3 Sheets-Sheet 3 Fig. 28 Steam Condensate NH; C03 H 0 zoiL Ursa t- H; 0/63 Attorney United States Patent 3,579,636 UREA SYNTHESIS PROCESS Ivo Mavrovic, 530 E. 72nd St., New York, N.Y. 10021 Filed May 16, 1968, Ser. No. 729,764 Int. Cl. C07c 127/00 U.S. Cl. 260555 11 Claims

(continued)

Table 3.6 (continued)

Urea synthesis process: the patent summary update				
Ref.	Patent name and no.	Year published	Description of patent	Patent no and details
[49]	Method for preparation of ammonia gas and CO_2 for a urea synthesis process US 20170210703 A1	Jul 27, 2017	The invention relates to a process for preparing ammonia gas and CO_2 for urea synthesis. In the process of the invention, a process gas containing nitrogen, hydrogen, and carbon dioxide as main components is produced from a metallurgical gas. The metallurgical gas consists of blast furnace gas, or contains blast furnace gas at least as a mixing component. The process gas is fractionated to give a gas stream containing the CO_2 component and a gas mixture consisting primarily of N_2 and H_2. An ammonia gas suitable for the urea synthesis is produced from the gas mixture by means of ammonia synthesis. CO_2 is branched off from the CO_2-containing gas stream in a purity and amount suitable for the urea synthesis	Inventors: Christoph Meißner, Denis Krotov, Olaf Von Morstein, Matthias Patrick Krüger Original Assignee: Thyssenkrupp Ag

References

1. Smil, V. (2004). *Enriching the earth: Fritz Haber, Carl Bosch, and the transformation of world food production.* MIT press.
2. Maxwell, G. R. (2004). *Synthetic nitrogen products: a practical guide to the products and processes,* pp. 267–284.
3. Meessen, J. H., & Petersen, H. (1996). *Ullmann's encyclopedia of industrial chemistry.*
4. Isla, M. A., Irazoqui, H. A., & Genoud, C. M. (1993). Simulation of a urea synthesis reactor. 1. Thermodynamic framework. *Industrial and Engineering Chemistry Research, 32,* 2662–2670.
5. Irazoqui, H. A., Isla, M. A., & Genoud, C. M. (1993). Simulation of a urea synthesis reactor. 2. Reactor model. *Industrial and Engineering Chemistry Research, 32,* 2671–2680.
6. Jiles, D. (2015). *Introduction to magnetism and magnetic materials.* CRC press.
7. Alqasem, B., Yahya, N., Soleimani, H., Qureshi, S., Irfan, M., & Rehman, Z. U. (2016). Experimental study on the effect of saturation magnetization of hematite nanocatalyst for green ammonia production. In: *AIP conference proceedings*, p. 030002.

8. Kiebarev, S., Ea, P., & Shustov, A. (1972). Influence of a magnetic-field on the photooxidation rate of some crystalline aromatic-hydrocarbons. *Doklady Akademii Nauk SSSR, 204,* 376.
9. Lawler, R., & Evans, G. (1971). *Some chemical consequences of magnetic interactions in radical pairs*. Providence: Brown University.
10. Frankevich, E., & Sokolik, I. (1971). Influence of magnetic field on the rate of photooxidation of aromatic hydrocarbons. *ZhETF Pisma Redaktsiiu, 14,* 577.
11. Turro, N. J. (1983). Influence of nuclear spin on chemical reactions: Magnetic isotope and magnetic field effects (a review). *Proceedings of the National Academy of Sciences, 80,* 609–621.
12. Schulten, K. (1982). Magnetic field effects in chemistry and biology. *Festkörperprobleme, 22,* 61–83.
13. Hayashi, H. (2004). *Introduction to dynamic spin chemistry: magnetic field effects on chemical and biochemical reactions,* vol. 8: World Scientific Publishing Co Inc.
14. McLauchlan, K. (1981). The effects of magnetic fields on chemical reactions. *Science Progress (1933-), 67,* 509–529.
15. Timmel, C., Till, U., Brocklehurst, B., McLauchlan, K., & Hore, P. (1998). Effects of weak magnetic fields on free radical recombination reactions. *Molecular Physics, 95,* 71–89.
16. Steiner, U., & Ulrich, T. (1989). Magnetic field effects in chemical kinetics and related phenomena. *Chemical Reviews, 89,* 51–147.
17. Yue, Q., Shao, Z., Chang, S., & Li, J. (2013). Adsorption of gas molecules on monolayer MoS 2 and effect of applied electric field. *Nanoscale Research Letters, 8,* 425.
18. Woodward, J. (2002). Radical pairs in solution. *Progress in Reaction Kinetics and Mechanism, 27,* 165–207.
19. Stass, D., Woodward, J., Timmel, C., Hore, P., & McLauchlan, K. (2000). Radiofrequency magnetic field effects on chemical reaction yields. *Chemical Physics Letters, 329,* 15–22.
20. Woodward, J., Jackson, R., Timmel, C., Hore, P., & McLauchlan, K. (1997). Resonant radiofrequency magnetic field effects on a chemical reaction. *Chemical Physics Letters, 272,* 376–382.
21. Timmel, C., & Hore, P. (1996). Oscillating magnetic field effects on the yields of radical pair reactions. *Chemical Physics Letters, 257,* 401–408.
22. Yahya, N., Qureshi, S., Rehman, Z. U., Alqasem, B., Fai Kait, C. (2017). Green urea synthesis catalyzed by hematite nanowires in magnetic field. *Journal of Magnetism and Magnetic Materials, 428,* 469–480.
23. Alqasem, B., Yahya, N., Qureshi, S., Irfan, M., Rehman, Z. U., & Soleimani, H. (2017). The enhancement of the magnetic properties of α-Fe2O3 nanocatalyst using an external magnetic field for the production of green ammonia. *Materials Science and Engineering B, 217,* 49–62.
24. Yahya, N., Alqasem, B., Irfan, M., Qureshi, S., Rehman, Z. U., Shafie, A., et al. (2016). The effect of saturation magnetization of nanocatalyst and oscillating magnetic field for green urea synthesis. *Physica B: Condensed Matter, 507,* 95–106.
25. Yahya, N., Puspitasari, P. (2013). Y3Fe5O12 nanocatalyst for green ammonia production by using magnetic induction method. *Journal of Nano Research,* 131–137.
26. Qureshi, S., Yahya, N., Kait, C. F., Alqasem, B., & Irfan, M. (2017). Enhanced catalytic activity of α-Fe2O3 with the adsorption of gases for ammonia synthesis. In *Materials science forum*, pp. 15–18.
27. Boxer, S. G., Chidsey, C. E., & Roelofs, M. G. (1983). Magnetic field effects on reaction yields in the solid state: an example from photosynthetic reaction centers. *Annual Review of Physical Chemistry, 34,* 389–417.
28. Atkins, P., & Lambert, T. (1975). The effect of a magnetic field on chemical reactions. *Annual Reports on the Progress of Chemistry, Section A: Physical and Inorganic Chemistry, 72,* 67–88.
29. Sagdeev, R., Molin, Y. N., Salikhov, K., Leshina, T., Kamha, M., & Shein, S. (1973). Effects of magnetic field on chemical reactions. *Organic Magnetic Resonance, 5,* 603–605.

30. Jing-Biao, C., Feng-Zhi, W., Dong-Hai, Y., & Yi-Qiu, W. (2001). Alternating current zeeman and stark interference effect in ramsey separated oscillating fields. *Chinese Physics Letters, 18,* 202.
31. Autler, S. H., & Townes, C. H. (1955). Stark effect in rapidly varying fields. *Physical Review, 100,* 703.
32. Hedvall, J. A. (1935). Über die im Verlauf von Phasenäunderungen in festen stoffen auftretende erhöuhung der chemischen Reaktionsfähigkeit. *Zeitschrift für Elektrochemie und Angewandte Physikalische Chemie, 41,* 445–448.
33. Hedvall, J. A., Hedin, R., & Persson, O. (1934). Ferromagnetic transformation and catalytic activity. *Zeitschrift fur Physikalische Chemie, Abteilung B, 27,* 196.
34. Petrucci, H., & Madura, H. (2007). *General chemistry principles and modern applications.* New Jersey: Prentice Hall.
35. Levenspiel, O. (1999). *Chemical reaction engineering* (3rd ed.), pp. 211–214.
36. Scott Fogler, H. (2005). *Element of chemical reaction engineering* (4th ed.), pp. 511–512.
37. Kjeldahl, J. Z. (1883). A new method for the determination of nitrogen in organic bodies. *Analytical Chemistry* 22, 366.
38. Richardson, J. T. (2009). *Journal of Applied Catalyst.* 3, 1781–1786.
39. Mavrovic, I. (1965). Urea Synthesis Process. U.S. Patent 3,172,911, issued Mar 9, 1965.
40. Cook, L. H. (1974). Urea Synthesis Process. U.S. Patent 3,816,528, issued Jun 11, 1974.
41. Cook, L. H., & Mavrovic, I. (1965). Urea Synthesis Process. U.S. Patent 3,200,148, issued Aug 10, 1965.
42. Maxim, K. (1972). Urea Synthesis Process. U.S. Patent 3,668,250, issued Jun 6, 1972.
43. Marten, J. H., Nixon, N. J. (1967). Urea Synthesis Process. U.S. Patent 3,354,205 issued Nov 21, 1967.
44. Mavrovic, I. (1981). Urea Synthesis Process. U.S. Patent 4, 296,252, issued Oct 20, 1981.
45. Cook, L. H. (1968). Urea Synthesis Process. U.S. Patent 3,370,090, issued Feb 20, 1968.
46. Mavrovic, I. (1970). Urea Synthesis Process. U.S. Patent 3,527,799, issued Sep 8, 1970.
47. Sakata, E., & Kojima, Y. (2003). Urea Synthesis Process. U.S. Patent 6,518,457, issued Feb 11, 2003.
48. Mavrovic, I. (1971). Urea Synthesis Process. U.S. Patent 3,579,636, issued May 18, 1971.
49. Christoph, M., Denis, K., Olaf, V. M., & Matthias, K. P. (2017). Method for preparation of ammonia gas and CO_2 for a urea synthesis process. U.S. Patent 20,170,210,703 issued Jul 27, 2017.

Chapter 4
Efficacy of Green Urea for Sustainable Agriculture

4.1 Introduction

A recent survey concentrated on the impact of the Haber–Bosch process in the leading scientific journal Nature Geoscience stated that "*Haber-Bosch is perhaps the most significant invention of the 20th Century, yet it has many side effects. Now we need an invention that changes the world just as such, but without the environmental impact*" [1]. Based on this statement, Green Urea could create a breakthrough in urea–ammonia synthesis research, and the technical description in Chaps. 1 and 2 are able to provide significant information of this novel invention. To further enhance the efficacy of Green Urea in real-time applications, a strategic delivery system is essential.

Slow release fertilizer (SRF) and control release fertilizer (CRF) design and application have been made advance in parallel with the development of nanotechnology and material engineering research [2]. The prominent characteristic of these advanced fertilizers over the conventional method is the gradual nutrient release pattern that is able to reduce excessive fertilization and multiple applications in the field [2, 3]. This phenomenon has its advantages in agriculture production systems by ensuring efficient nutrient utilization by plants thus, minimizing environmental blueprints, and human health [2, 3] threats. This chapter focuses on the efficacy of Green Urea through integration with slow and control release fertilizers and their its effects on plants grown under controlled environment in nursery. With the population boom and the increasing demand for land to produce more food to support this growth, vertical farming known as an urban agriculture is also discussed as a possible solution to increase agriculture production [4]. The concept of vertical farming is relatively new, and is growing expoinentially. Vertical farming is a method that involves multiplying the area of farming by optimizing its height. Most modern cities nowadays use these techniques in order to put up with the market demand.

Integration of Green Urea with the latest technologies such as slow and control release properties integrated with vertical farming could potentially play a major

N. Yahya, *Green Urea*, Green Energy and Technology,
https://doi.org/10.1007/978-981-10-7578-0_4

part in improving nutrient efficacy in field applications. The combination of technologies and green ammonia synthesis would be able to highlight Green Urea as a role model to face sustainability issues in agriculture landscape.

4.1.1 Importance and Challenges of Nutrient Use Efficiency

The sustainability of crop production highly depends on the implementation of nutrient management practices. The nutrients supplied from fertilizers are essential for plant growth and are indirectly essential to the world's ecosystem. However, the objective to achieve optimum nutrient use efficiency (NUE) remains as a major challenge in agriculture production systems [5]. This has been steadily researched by agronomist and fertilizer technologist for over the past few decades to increase crop production and at the same time minimize potential hazards on human health and the environments.

Generally 17 nutrients for crop growth which include three from water and air (carbon, hydrogen, oxygen); three macronutrients (nitrogen, phosphorous, and potassium) mostly supplied from soils; secondary nutrients (sulfur, calcium, magnesium) usually needed in smaller amounts and micronutrients (boron, iron, manganese, zinc, copper, molybdenum, chlorine, nickel) needed in minor amounts but are vital for the metabolic process involved in crop development [5]. Factors that contribute to the use of nutrients in crop production are the interaction between climate, soil, plant, and the socio-economic status of the farmers [6].

Among the plant nutrients, nitrogen is considered the most limiting crop production nutrient in many part of the agriculture practicing areas the world [7]. However, the efficiency of nitrogen fertilizers uptake by the crops impoverished, whereby only less than 50% utilized by the crops [7]. A conventional way of nitrogenous fertilizer application caused excessive nitrogen losses through runoff, leaching, and volatilization [8, 9]. Runoff leads to eutrophication and leaching which is normally associated with high rainfall groundwater pollutions [9]. This ultimately limits the uptake of essential nutrients by the crops and consequently results in poor growth and yields [2, 3, 8, 9].

Therefore, in order to develop strategic fertilizer management practices the 4R principle has been endorsed by the International Fertilizer Industry which consists of using right fertilizer, in the right dose, at the right place, and at the right time [9]. Besides that manipulating abiotic and biotic conditions and the introduction of crops with genotypes within the same species and high nutrient efficiency are also being initiated to increase efficient use of fertilizers. Although it is a challenging task, the maximum improved nutrient efficiency will benefit the economy tremendously, for both farmers and consumers with increased yield, lower environmental pollutions, and lower operational cost [9]. In the next section of this chapter, slow release (SRF) and controlled-release (CRF) fertilizer application in agriculture systems as well as its properties as an integration with Green Urea to increase its efficiency is discussed. A summary on the techniques that can be implemented to improve nitrogen use efficiency is illustrated in Table 4.1.

Table 4.1 Technologies to improve nitrogen use efficiency [10]

Technology	Source of nitrogen	Techniques	Technical application
Slow/ Control release	Amide nitrogen, Ammonium nitrogen	PSU/SCU/PCSU/ NCU	Gradual release of nitrogen into the soil via physical coating on the surface of urea
		MU/IBDU/UF/CDU	Delays the decomposition rate of urea in soil via chemical condensation reaction between urea and aldehyde
		Chelate urea by organic matters	Chelate urea by organic matters like humic acid to delay the decomposition rate of chelate urea in soil
Urease inhibitor	Amide nitrogen	NPBT/NPPT	Slows down conversion of urea decomposition into ammonium nitrogen by inhibiting urease activity in the soil
Nitrification inhibitor	Amide nitrogen, Ammonium nitrogen	DCD/DMPP/ Nitrapyrin	Slows down the conversion of ammonium nitrogen by inhibiting the activity of nitrification bacteria in soil
Fertilization	Nitrate nitrogen, Ammonium nitrogen, Amide nitrogen	Deep application of nitrogen fertilizer	Minimize volatilization of nitrogen fertilizer in the form of NH_3
		Balance the ratio of water and fertilizer application	Water and fertilizer are applied according to the demand of various crops in different growth stages to increase utilization rate of fertilizer
		Balance the ratio of nitrogen and other plant nutrients	Macro, secondary, and micronutrients are applied in proper manner to enhance synergetic relationship in nutrient elements
		Split fertilization	Applied at different stages to reduce the loss of plant at one time in fertilization
Biological nitrogen fixation	N_2 from the atmosphere	Rhizobiaceae	Manipulating nitrogen-fixing microorganism to fix nitrogen in the air to increase the amount of ammonium nitrogen supply for the crops
Plant breeding	Nitrate nitrogen, Ammonium nitrogen	Transgenic breeding, Cross-breeding	To develop transgenic crops through biotechnology to increase uptake of N (example: rice and corn)

4.2 Slow Release and Control Release Fertilizer

4.2.1 Definition and Characteristic of Slow/Control Release Fertilizer

One of the significant progresses in fertilizer research that could increase NUS is the development of slow release (SRF) and control release fertilizer (CRF). The development has been initiated the 1950s and became prominent during the 1990s. According to The Association of American Plant Food Control Officials, CRF and SRF are fertilizers that consist of nutrients in a form which delay or increase the availability period in synchrony with the plants need [11].

The term for CRF and SRF was distinguished whereby CRF is manifested by the factors of rate, pattern, and duration period of release which is well known and

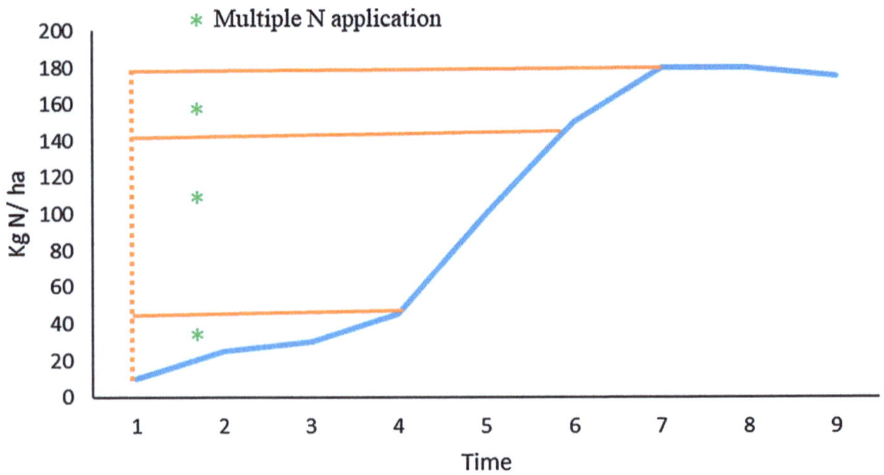

Fig. 4.1 "Enhanced-efficiency fertilization concept" based on multiple N application system [15]

controllable during the preparation while SRF releases the nutrients at a slower rate but does not dependent on the factors discussed for CRF [12]. Thus, SRF is more fragile toward storage conditions, transportation, soil conditions, and biological activities [13]. The release pattern, quantity, and duration of CRF can be predicted over SRF in field applications [14]. The gradual releases of nutrient which regulates according to the need of the plant and elimination of conventional practice of constant fertilization, is the prominent characteristics of these fertilizers.

The sigmoid (S-Shape) pattern relationship known as "enhanced-efficiency fertilization concept" is a relationship between nutrient demand with the availability of fertilizers and is illustrated in Fig. 4.1 [15]. It is accepted that the fertilizers should release the nutrients in sigmoid pattern to optimize the uptake of plant nutrient and reduces overall negative environmental impact. The concept benefits agriculture production; However, applying multiple or split technique application of N fertilizer is labor extensive and causes additional operational cost. Therefore, most farmers prefer one application of controlled-release fertilizers over multiple applications with conventional fertilizers. In Japan, the development of controlled-release fertilizers led to the innovation of agro-technologies for various field crops including new concepts of fertilizer applications [16].

4.2.2 Development and Benefits of Slow/Control Release Fertilizer

Four important elements for approaching CRF and SRF are (i) stabilizing agents for controlling the conversion of nutrients in soil such as urea inhibitors (NBPT) and nitrification inhibitors (DMPP, Nitropynin, DCD), (ii) development of compounds

Table 4.2 Advantages and limitation of coating materials for control and slow release fertilizers [8]

Material	Advantages/drawbacks	General disadvantages [8]
Sulfur-coated urea [18, 19]	• Low cost • Not complicated process • Relatively fast compared to polymer-coated urea • Unpredictable release	Susceptible toward: • Heat and light • Mechanical forces • Poor sealing • Soil properties • Complicated mechanism • Lengthy process • Relatively expensive • Partially effected by soil properties • Highly expensive • Mostly non-degradable • Pollutant • Toxic
Sulfur-polymer-coated urea [20]	• Double-layered coating • Slow and regular release • Better sealing • Least sensitive to heat and rough handling compared to coating without polymer	
Polymer-coated urea	• Synthetic product • Hydrophobic • Simple mechanism • Least affected by soil properties • Completely controlled release	
Starch base polymer-coated urea [8]	• Generic product • Abundantly available • Low cost • Biodegradable • Environmental friendly	• Weak coating barrier • Hydrophilic • Need modifications

with less solubility, (iii) development of compounds need to be degraded before becoming available for plants, and (iv) coating technology with physical barrier control release [9].

In the current time frame, great attention is given to coating technology. There are generally two classes of coated CRF; namely (a) sulfur-coated and (b) polymer-coated [8, 9]. Sulfur-coated fertilizers were developed during the 1960s, and it is selected as coating material due their low costs and providing secondary nutrients after the oxidation process [8, 9]. Moreover, lately studies were conducted by using natural coating agents and nanocomposites to improve the coating techniques of fertilizers. The thickness and types of materials are the factors that determine the efficacy of coating properties of CRF. Various coating materials and their advantages and limitations are shown in Table 4.2.

Fig. 4.2 Leaf growth of spinach (45 days old seedlings) fertilized with chitosan-coated controlled-release Green Urea (left) and with conventional nitrogen fertilizer (right). Spinach fertilized with conventional fertilizer shows toxicity effect of conventional fertilizer

These fertilizers have a significant impact on the reduction of nutrient loss, enhance NUE (sigmoidal release patterns), and minimize environmental pollution. Toxicity of young plants or seedlings among the phenomena that occurs due to the application of soluble conventional fertilizers. However, with strategic application of CRF, high ion concentration of salt and osmotic stress that could damage the plant at the seedling stage could be avoided. One of the examples that shows the toxicity effect of soluble conventional fertilizers is shown in Fig. 4.2. Although slight acidification is noted the application of sulfur-coated control release fertilizer, this condition favors the uptake of phosphorus and iron of plants from the soil. Single application benefits the farmers in term labor saving, energy, and operational cost [16]. The prediction of nutrient release can be performed particularly for polymer coated fertilizers or with hydrophobic materials attributable to their lower level of sensitivity to soil and climatic conditions. This will substantially enable reliable software programmes to investigate the relationship between the climatic condition, release pattern, and plant growth [17].

4.2.3 Mechanism of Control Release Fertilizer

The mechanism of nutrient release from CRF has been well described in many literatures [17, 21]. The water as vapor passes through the coating process condenses and dissolves the fertilizer core by triggering internal pressure and inducing two distinguished event to take place. First, the coating breaks and releases the entire content of the granule with immediate effect [17]. This mechanism is called “failure mechanism” [22] which is usually related to inorganic types of sulfur

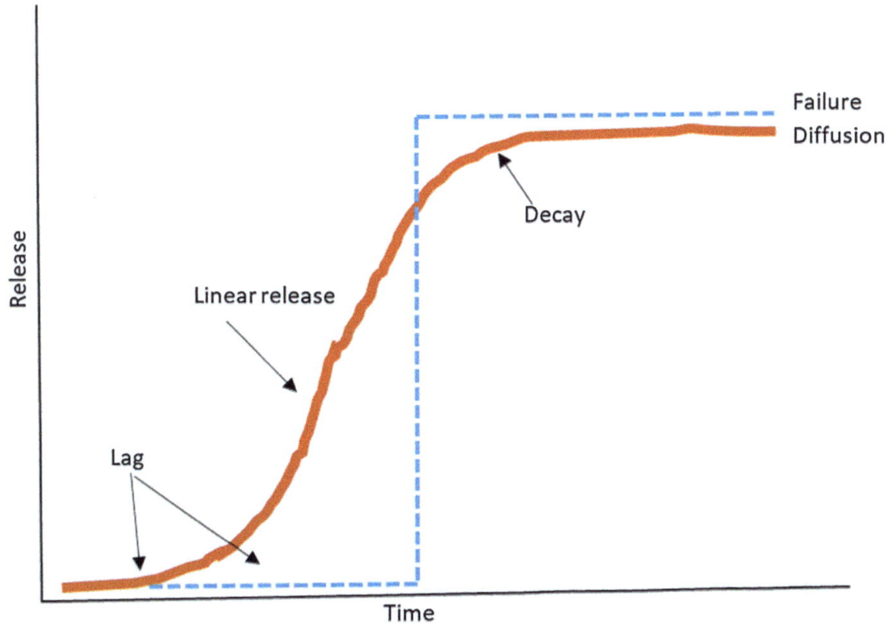

Fig. 4.3 Release pattern release-coated urea showing the stages of diffusion vs failure (Failure is without slow or control release) [17]

coating which are fragile and non-elastic. However, if the coatings resist the pressure, their release factor is driven by diffusion force by a concentration gradient or mass flow, or both in combination [17]. This mechanism is clearly explained in Fig. 4.3. This is usually related to polyolefin and resin coatings [17]. Three stages of nutrient release which have been described are (i) the initial stage of "lag period" with limited or almost no release, (ii) the constant release stage, and (iii) the mature stage whereby a gradual reduction in the release rate is observed [21] (Fig. 4.4).

4.2.4 Economic Outlook of Slow/Control Release Fertilizer

The world's consumption of slow and controll-release fertilizers (metric tonnes of fertilizer product) is shown in Table 4.3. After the year 2005, a sharp increase in the consumption of these fertilizers was observed in China due to the large scale of production of sulfur-coated urea and active development of polymer-coated fertilizers. There were no cost studies conducted for these fertilizers. Based on Fig. 4.5, prices for slow and control release fertilizers are higher compared to conventional fertilizers. However, the cost gap can be minimized if these fertilizers efficiency are emphasized [12].

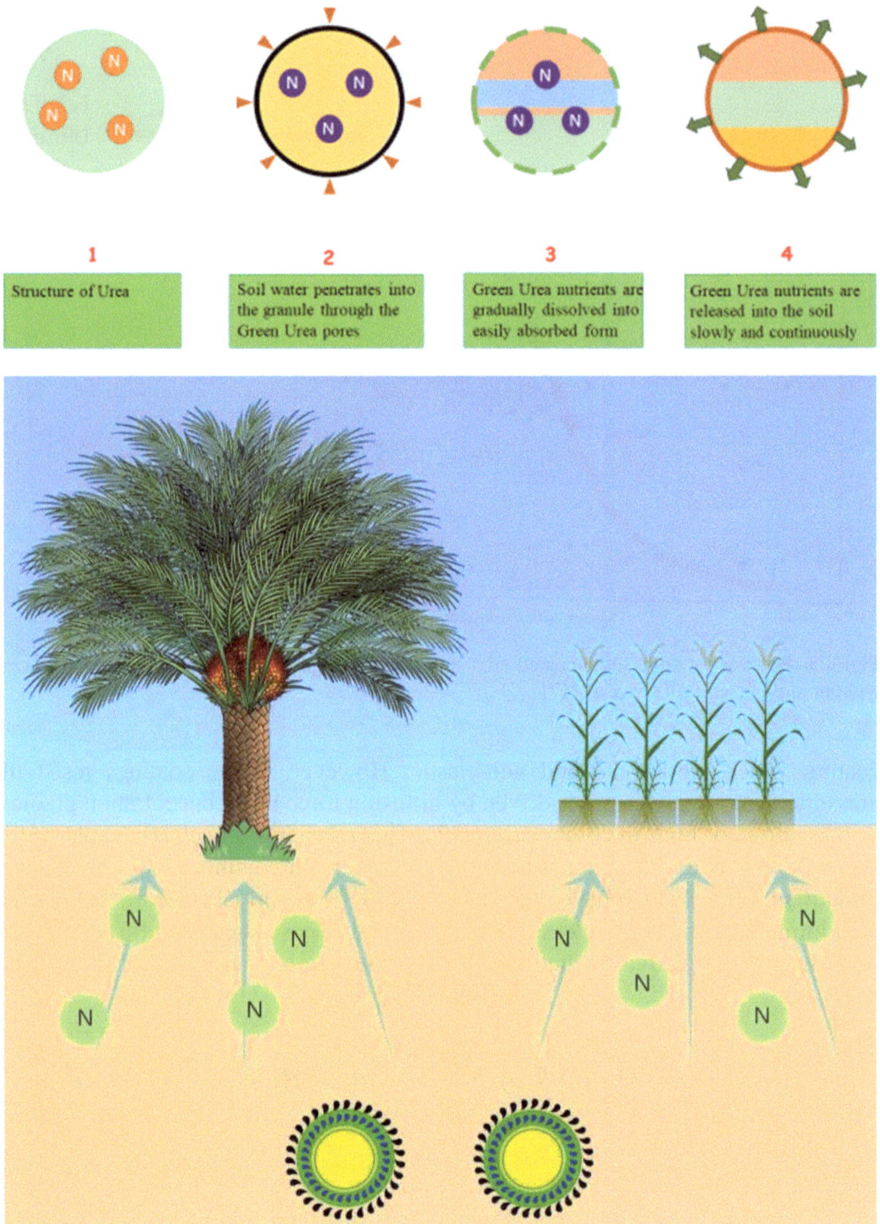

Fig. 4.4 Mechanism of Green Urea fertilizer

Table 4.3 World consumption for slow release/control release fertilizer [12]

Region	1983	1995/96	2004/05	2006/07 Estimate
United States	202,000	357,000	569,000	590,000
Western Europe	76,000	87,000	120,000	125,000
Japan	47,000	96,000	97,000	110,000
China	–	–	–	1,350,000
Canada	–	–	–	150,000
Total	325,000	540,000	786,000	2,275,000

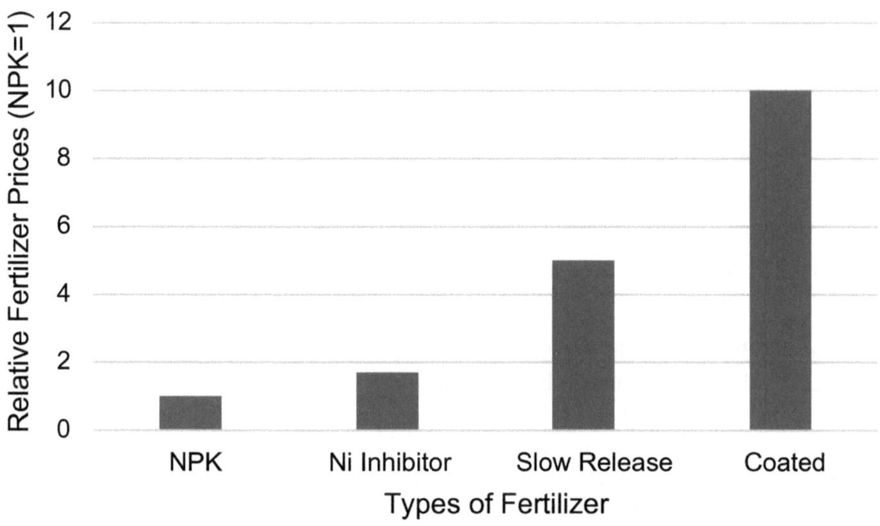

Fig. 4.5 Comparison of relative fertilizer prices between slow and control release, conventional, and standard fertilizers [15]

4.3 Slow Release Urea Acid

Green Urea was initially defined by the synthesis method which requires thermal energy and pressure. Green urea was confirmed in the acetonitrile (ACN) solution using FTIR spectroscopy. Further experiments were performed to synthesize urea acid for the slow release properties. Urea granules are produced by using temperature to convert molten urea to solid urea. These granules are coated with geopolymer coating and urea sulfate produced using sulfuric acid. FTIR spectroscopy of solid urea produced in the present research and commercial urea shown in Fig. 4.6.

The stretching of the N–H bond that is a characteristic band of primary amine was observed. This band is composed of two peaks which correspond to each vibration of the N–H bond of primary amine: For commercial urea, one is at 3434 cm^{-1} and the

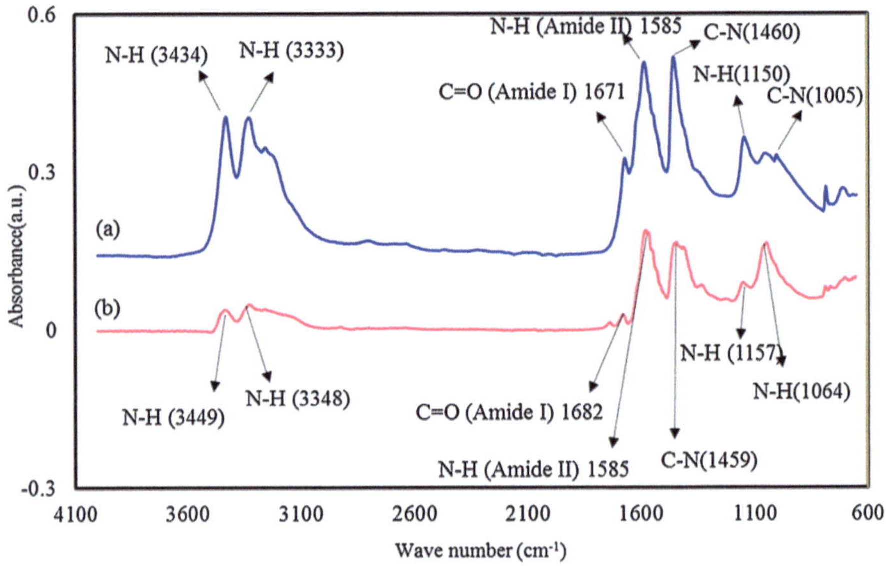

Fig. 4.6 FTIR spectra of **a** Green urea and **b** Commercial urea

other at 3333 cm^{-1} while experimmental Green urea peaks are at 3449 cm^{-1} and 3348 cm^{-1}. The N–H bond in both spectra is in the same range. Another band has appeared in commercial urea and the Green urea at 1671 cm^{-1} and 1682 cm^{-1}, respectively, for the C = O group. A band is observed at 1460 cm^{-1} and 1459 cm^{-1} attributed to the vibrational stretching of the C–N bond of urea [23, 24].

The comparison of FTIR spectra of commercial urea and Green Urea confirms the peaks of Green Urea in FTIR spectra. The urea is soaked in a solution of sulfuric acid and sulfuric acid plus chitosan to produce ammonium sulfate. Chitosan is a natural biopolymer modified from chitin. This is the main structural component of squid pens, cell walls of some fungi, shrimp, and crab shells. Chitin and chitosan are copolymers that are found together in nature. The specific properties of these copolymers are that they are environmentally friendly and easily degradable. Chitosan is a linear polysaccharide composed of randomly distributed β-(1 → 4), D-glucosamine (deacetylated unit), and N-Acetyl-D-glucosamine (acetylated unit). It is prepared by treating the chitin shells of shrimp and other crustaceans with an alkaline substance, such as sodium hydroxide [25]. Figure 4.7 shows the spectra of Green urea, geopolymer-coated urea, urea sulfate, and chitosan with urea sulfates produced using the Green urea.

Green urea granules are also coated with geopolymer. Geopolymer is a material that involves naturally occurring silico-aluminates. It is based on the chemistry of alkali activated materials. Geopolymer is a potential material to be developed as urea coating material. It exhibiting good mechanical property which is required to avoid coatings from been ruptured easily during storage and handling inorganic binders [2]. The corresponding FTIR spectra together with pure Green urea are

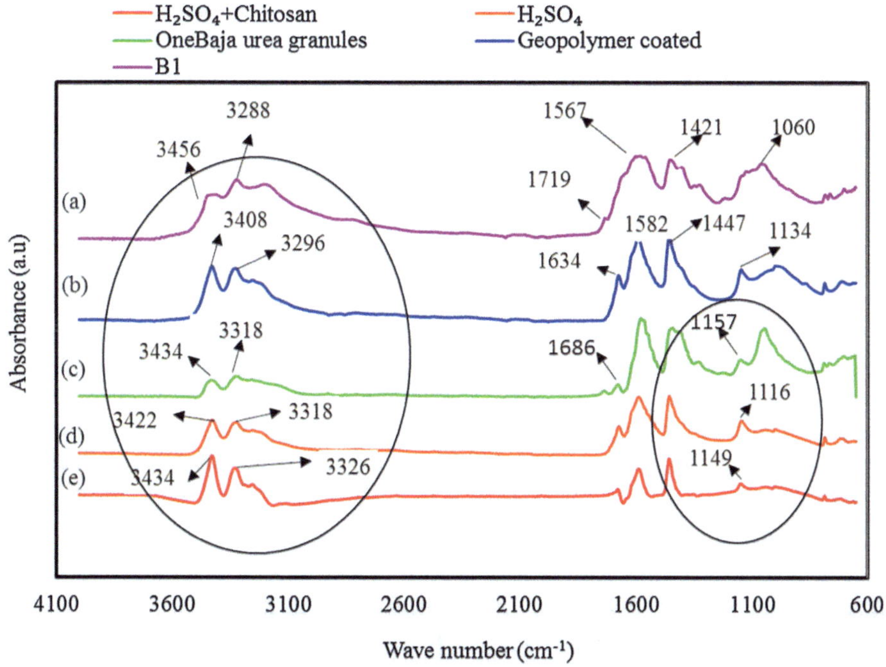

Fig. 4.7 FTIR spectra of (B1), (b) Geopolymer-coated urea, (c) Green urea, (d) Green Urea plus sulfuric acid, (e) Green Urea plus sulfuric acid and chitosan

presented in Fig. 4.7. In region A, ammonium band (2400–3520) cm^{-1} is difficult to differentiate from the urea peak. The only difference that can be seen is in term of height and a peak shift from 3318 cm^{-1} to 3326 cm^{-1} in the FTIR spectra of Green urea with sulfuric acid plus chitosan. The best ammonium sulfate FTIR peak assignment is at 1116 cm^{-1} [26].

In this study, the 1116 cm^{-1} peak in the FTIR spectra of sulfuric acid plus Green urea can be attributed as ammonium sulfate peak. In the region B, the shift in the Green Urea peak, 1157 cm^{-1}, can be observed, for sulfuric acid plus Green Urea the peak that is at 1116 cm^{-1}, for urea with sulfuric acid and chitosan peak is shifted to lower frequency as compared to pure urea but this shift could be the presence of chitosan. Hence, the FTIR result proved the formation of ammonium sulfate in the sample of urea plus sulfuric acid.

FTIR spectra of geopolymer-coated urea are shown in Fig. 4.7b. The comparison between pure Green Urea and geopolymer-coated urea shows that the peaks in the region of (3400–3200) cm^{-1} have been shifted to lower frequencies. The characteristic peak of pure Green Urea has also shifted from 1686 cm^{-1} to 1634 cm^{-1}. The shift in the FTIR peaks for geopolymer-coated urea is due to the presence of geopolymer.

FTIR of sample B1 is shown in. In this sample, peaks in the region of (3400–3200) cm^{-1} have been shifted and characteristic peak of ammonium peak appeared at 1421 cm^{-1} [27]. Overall, in this sample, peaks are broad and the Green Urea peak at 1686 cm^{-1} has shifted to higher frequency at 1719 cm^{-1}. It can be seen in the Fig. 4.7 that the overall spectra are dominated by Green Urea peak. The C = O in urea is longer than the one found in the carbonyl group. The C–N band is shorter than the amines and amides. The peak shift in FTIR spectra is due to the different characteristics infrared bands. Each peak in the FTIR spectra corresponds to a specific band. The frequency, intensity, peak width, and shift of any particular IR vibration are sensitive. Any change in the Green Urea can cause the change in the IR spectra. The number of peaks present at various frequencies provides the information about the corresponding molecules. Peaks of ammonium sulfate, geopolymer, B1, sulfuric acid, and sulfuric acid plus chitosan are identified on the basis of their molecular structure and orientation. Molecular structures are composed of negative or positive ions. Theses ions have a significant effect on the frequency shift of the samples discussed above. The charge and ions of various radii generate an electric field in the salts. These electric fields affect the vibrational frequencies.

4.4 Case Study: Efficacy of Coated Green Urea on Vegetative Plant Growth of Spinach

About 20–70% of conventional urea applied in the fields or plantation is loss to the environment through leaching and volatilization [8]. Thus, the Green Urea must be manifested as an EEF to improve its effectiveness by providing efficient nutrient supplies to the crops and maintaining an as eco-friendly fertilizer. In this section, the coating procedures and dissolution test conducted on Green Urea are discussed and followed by the highlights on the evaluation on its efficacy on spinach growth under nursery condition.

4.4.1 Coating and Dissolution Test of Green Urea

Green Urea coating process initiates by placing the fertilizers in the fluidized bed coater on gas distributor plate. The schematic diagram is illustrated in Fig. 4.6. Air is then blown from the bottom into the fluidized bed by a high-pressure blower while the coating material is sprayed from the upper side of the bed. The air would pass through the air heater before moving upward through the gas distributor plate making the urea granules fluidize. The blown hot air from the bottom would be sucked from the top of the bed using an air from the top using an air suction pump installed at the top of the bed reactor (Fig. 4.8).

The coating process of urea of each batch takes nearly 15 min. The experiments were carried out on four samples of coating materials namely (i) zeolite, (ii) carbon

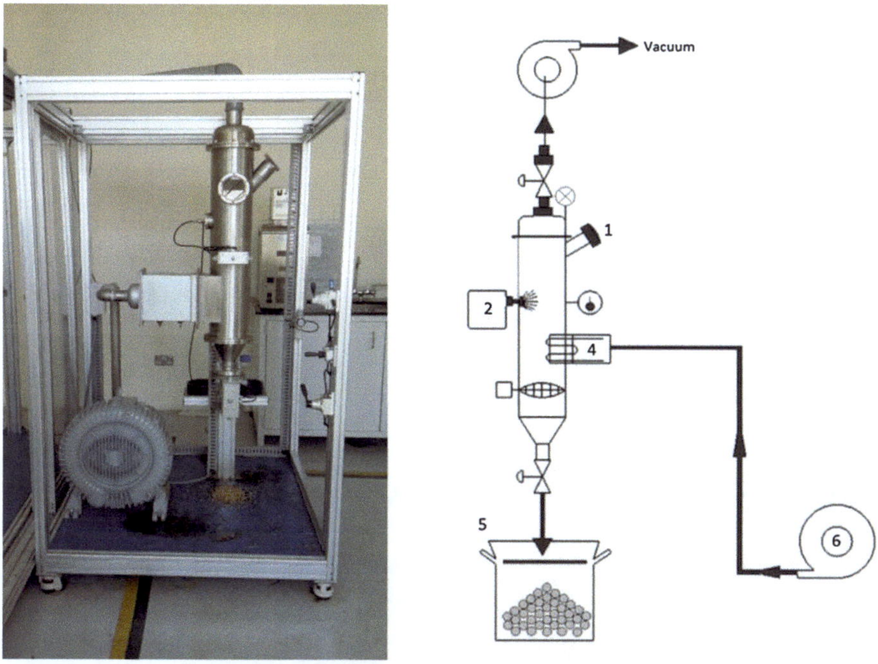

Description of the coating urea reactor	
1	Uncoated Urea Inlet
2	Coating Sprayer
3	Gas Distributor Plate
4	Air Heater
5	Coated Urea
6	High Pressure Blower

Fig. 4.8 Schematic diagram of the fluidized bed reactor

structure, (iii) thin coating of chitosan, and (iv) thick coating of chitosan. Following the coating process of urea, control release time of uncoated and coated urea was examined. 2 g of each urea samples was placed inside a beaker filled at 10 ml of distilled water. The beaker was then placed on a magnetic stirrer plate with speed of 120 rpm. The release time was then calculated for each sample.

The coating material from zeolite, carbon structure, and two kinds of coating (thick and thin) from chitosan biopolymer material were used on the granulated Green Urea. The relation between the percentage of urea released and the time taken for the dissolution of each sample of urea is tested. The results demonstrate that the longest time taken for urea sample dissolution is the thick coating of chitosan 7 min followed by carbon structure material 6 min. Thin coating chitosan

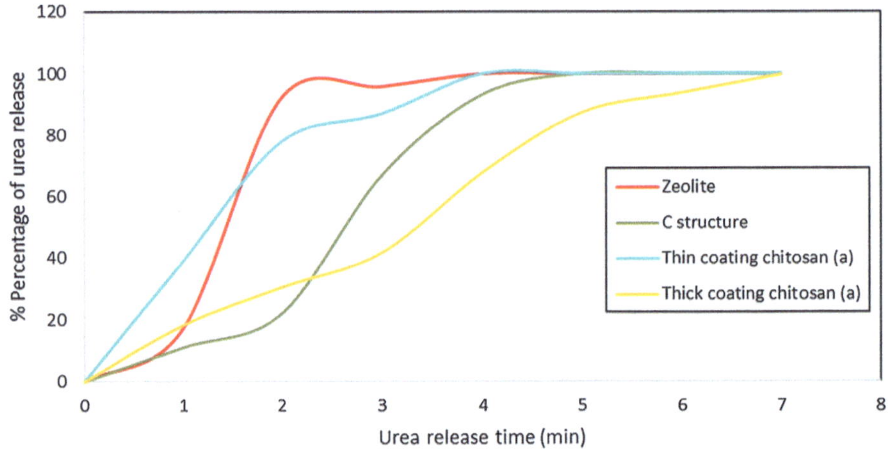

Fig. 4.9 Percentage of urea release for four different properties of coating materials

Fig. 4.10 **a** Coated thick-coated chitosan material and **b** Green Urea with ÓneBAJA' trademark packed in bottles for further testing in fields

and zeolite took 5 min to dissolve (Fig. 4.9). Based on these results, thick coating chitosan was selected for subsequent plant experiments the nursery. The coated thick coating chitosan is shown in Fig. 4.10.

4.4.2 Response of Spinach Plants to Controlled-Release-Coated Green Urea (CRCGU)

Spinach (*Spinacia oleracea*) under the family member of Chenopodiaceae is among the popular leafy vegetable attributable to its nutritional value. It has high content of

Table 4.4 Description of treatments employed in the experimental study

Treatment	Description	Time of application
T1	Control without any fertilizer	Interval of 10 days
T2	90 kg N/ha conventional fertilizer	Interval of 10 days
T3	5% (w/v) chitosan as foliar spray	Interval of 10 days
T4	90 kg N/ha CRCGU (0.3 g per pot)	Interval of 10 days
T5	Combination of 5% (w/v) chitosan as foliar spray + 90 kg N/ha (0.3 g per pot) CRCGU	Interval of 10 days

vitamins such as vitamin A, ascorbic acid, riboflavin, thiamine, and appreciable quantities of minerals such as iron and calcium [28, 29]. In Malaysia, spinach is grown in lowland and occupies about 2474 ha of total area planted with vegetables (44,000 ha) [30]. The world's production of this leafy vegetable is dominated by China, followed by USA, and Japan [31].

Numerous studies have showed great improvement in yield and fertilizer use efficiencies through the interaction of control release urea in various agricultural system and crops. Nevertheless, its information on the effect of vegetable crops particularly spinach is limited. Hence, a controlled environment experiment under nursery condition was laid in a completely randomized design to determine the vegetative growth of spinach after the application of CRCGU and plant enhancer (chitosan). The experiment was conducted at Semaian Seri Iskandar, Bandar Sri Iskandar, and Perak. Commercial spinach seeds (green round variety) were obtained locally from Kayegaki Sdn Bhd, reputable agriculture company in Malaysia. The treatments were done as shown in Table 4.4.

CRCGU was applied as basal application by mixing with soil while chitosan was mixed with water to achieve 5% concentration and then sprayed directly on the leaves [43]. Garden pots were filled with red and black sieved soil. The soils were treated with the respective fertilizers before the seeds were sown in the pot. Six seeds were sown in rows about 5 cm apart from each other. The pots were watered twice daily and manually weeded manually on regular basis. The experiment was repeated twice using the same soil. Vegetative growth parameters were recorded and comprises leaf area, number of leaves, and average of leaf weight. Statistical analysis was performed on the collected data for each treatment by using analysis of variance (ANOVA) with a significant difference at 5% level between the treatments.

Total number of seedling germinated was the maximum with T2 in the experiment, respectively, after 5 days sowing period. However, all of treatments were not significant different compared to the control (T1). It is concluded that the fertilizers have less or no effect on the germination of spinach seeds (Fig. 4.11).

Figure 4.12 shows that on the 15th day of the experiment, the T5 treated plant recorded the highest plant height by achieving 10.59 cm, followed by T4 treated plants 8.53 cm. The highest plant height was recorded for T5 treated plants (23.63 cm) on the 30th day. By the end of the experiment, on the 60th day, T5 treated plants recorded the highest height 35.37 cm of both T4 and T5 were found

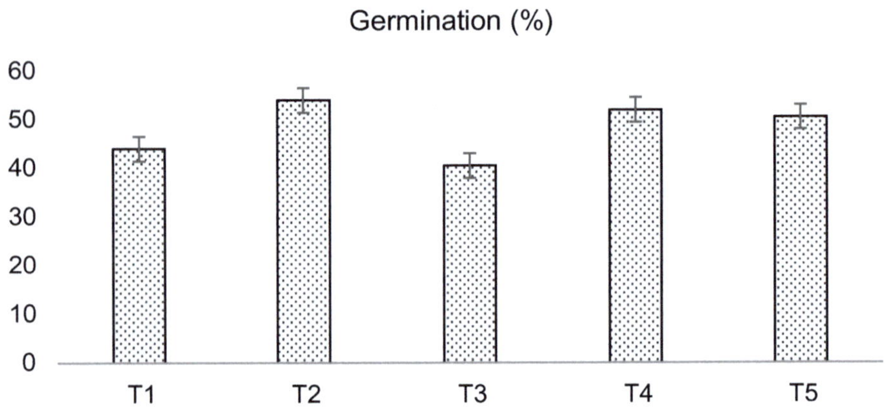

Fig. 4.11 Effect of control release Green Urea (CRCGU) on germination in spinach

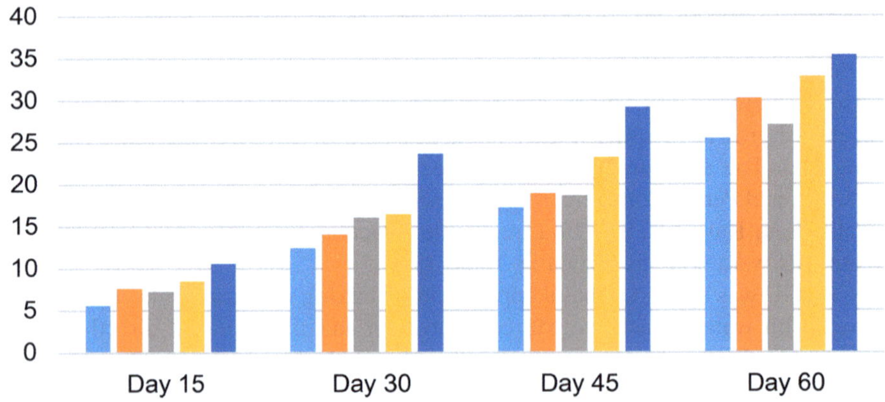

Fig. 4.12 Effect of control release Green Urea on plant height of spinach during 60 days of experimental period

significantly superior over T1, T2, and T3 at the 5% level of significance at 45th and 60th day of the experiment.

Figure 4.13 indicates that, produced significantly larger leaf area (compared to other treatments (average: 26.66 cm^2). A significant difference in leaf area was observed between T2 (conventional urea). It was found that the vegetative growth of spinach responded well with the combination of Green Urea and chitosan as foliar spray and T5, which exhibited the role of chitosan as an inducer for enhancing better plant growth.

Fig. 4.13 Effect of control release Green Urea on leaf area of spinach

4.4.3 Spinach Plant Tissue Test for Uptake of Nitrogen

Soil testing can provide an estimate of plant nutrient availability in soil. However, soil testing cannot predict the quantity of nutrients a plant or crop will actually use. Only through plant tissue analysis, can we assess the plant's nutritional status and determine how well the soil supplies the plant's nutritional requirements. Plant tissue analysis cannot replace a good soil testing program; However, plant tissue analysis can provide additional information on plant nutrient status not obtained from soil analysis. To test the urea produced by magnetic induction method, spinach is grown using water without any urea, commercial urea, and urea produced by the current method. Figure 4.14 presents the spinach plant grown by water (T0R1), experimental urea (T1R1), and commercial urea (T2R1). It can be seen in Fig. 4.15. The spinach grown with experimental urea is nourished and looks healthier as compared to the other two plants. The nitrogen (N) uptake concentration in the spinach plant tissue was performed for both experimental urea and commercial urea.

Fig. 4.14 Spinach plant **a** water, **b** experimental urea, **c** commercial urea

Fig. 4.15 Spinach plant grown from experimental urea

Table 4.5 N concentration in spinach plant tissue

Treatment	%N in tissue	Mean % N in tissue
Commercial urea R1	3.76	4.79
Commercial urea R2	5.53	
Commercial urea R3	5.08	
Experimental urea R1	3.15	4.12
Experimental urea R2	4.21	
Experimental urea R3	5.01	
Control released urea R2	3.03	2.79
Control released urea R1	3.23	
Control released urea R3	2.13	

The nitrogen concentration test was performed on spinach tissue for urea, control released urea, and experimental urea. Figure 4.16 shows the graph for mean uptake of nitrogen in spinach plant tissue.

Table 4.5 summarizes the results of N concentration in spinach tissue. The percentage of N for commercial urea is 4.79, experimental urea is 4.12, and for control released urea it is 2.79. The results demonstrate that the improvement in

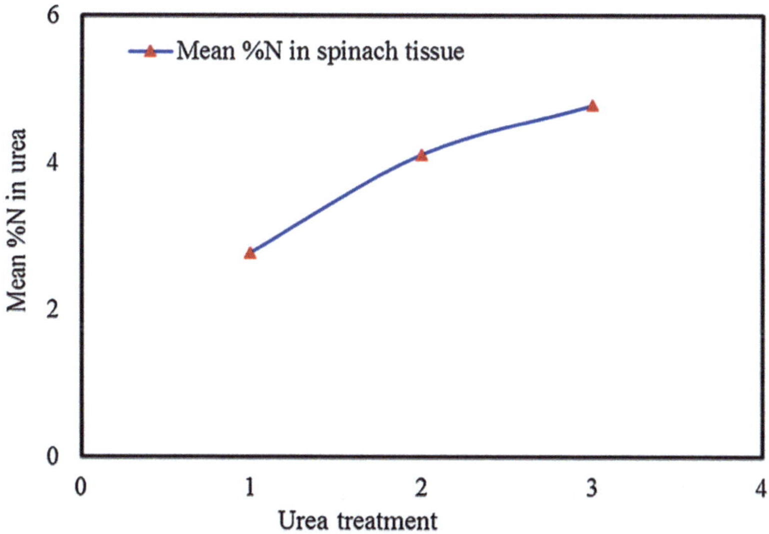

Fig. 4.16 Percentage uptake of nitrogen in spinach tissue

synthesis method of experimental urea can produce urea with high percentage of N (Fig. 4.16).

According to the results of ammonia and urea synthesis under applied magnetic field, it is obvious that the magnetic nanocatalyst along with applied magnetic field is not the only parameter that has an effect on chemical reaction. Some other parameters also have a significant effect on urea yield. Experimental urea can be further improved further to have higher content of nitrogen. In the following sections, flow rate of gases and time effect discussed on urea yield.

4.5 Vertical Farming

4.5.1 Overview of Vertical Farming

One of the approach to attain nutrient use efficiency in crop production system is the emphasis on effective management and the introduction of vertical farming. Vertical farming stacking of the farms small land can be utilized for more production. It utilizes less water, and the production is consistent because of the even exposure of the sunlight. Urban populations have shown great interest in these methods of farming as the food demand can be met, reduces cost of transportation, and environmental effects. Increasing per capita income in developing nations, and changes in occupational, and international relations have modified food choices and preference. All these with the blooming of the population will definitely change the agriculture industry.

Intensive agriculture has foreseen the rise in trend of inorganic agriculture, vertical farming to accommodate the demand the increasing world population, and environmental blueprints. It helps to produce in a faster and more especially for urban population. Vertical farming concept was given by Professor Despommier [32]. Vertical farming defined as a system of commercial farming whereby plants, animals, fungi, and other life forms are cultivated for food, fuel, fiber, and other products and services by artificially stacking them vertically above each other. Vertical farming is considered a large-scale agriculture in urban cities. The concept is integrated for the cultivation of fruits, vegetables, medicinal, fuel producing plants, and other plant products in cities [33]. Other benefits include reducing transportation and operational cost that come with conventional agriculture methods norm agriculture [34]. It is regarded as a way forward from greenhouses technology, because it involves the utilization of resources in various directions and subsequently meet the food particularly in cities Table (4.6).

Table 4.6 Three different types of vertical farming concept

Types of vertical farming	Descriptions
The utopian concept 	Utopian concept that displays the underground vertical farming as duplicated in Netherlands [35]
Open air sky scrapers 	Implemented in open air or by sky scrapers style for climate control and consumption. Developed for individual or community use. Not suitable for commercial purpose. Modification from this design involves cultivation of crops in the concept of sky scrapers by providing ambient amount of sunlight
Closed system sky scrapers 	Cultivation of plant in sky scrapers under closed system. Targeted for large-scale cultivation. These systems under trials at various locations (Singapore, Canada, London). An estimate of vertical farm of 9300 m^2 with 30 stories should provide around 15,000 people with 2000 kcal of nutrition per day [36]

4.5.2 *The Advantages of Vertical Farming*

The vital aspect or idea of vertical farming is an ecological farming concept [37, 38] which works well with nature. The management practice does not disrupt the environment nor its ecosystem. Due to this advantageous note, it is considered as an alternative for organic farming which is the most eco-friendly means of agricultural practice. Moreover, its advantage takes over chemical-based farming, whereby excessive use of chemical-based fertilizers and pesticides leads to eco-toxicity and health hazards [5, 37, 39]. Nutrient management practices like crop rotations, soil fertility building via nitrogen, and nutrient recycling using organic material such as crop residues, farmyard manure, and minimization of use of chemical-based fertilizers [38]. Control of pest populations relies on use of resistant crops, crop rotation, increase in predators for natural control of the pests, and increase in genetic diversity along with the judicious use of water resources and animal husbandry [39] (Table 4.7).

Table 4.7 Advantages of vertical farming

Advantages of vertical farming	Details
Increase in production and availability in crops	This technology produces sustainable crop production and is not dependent on environmental factors. As an example, a 30 storeys high building with a basal area of 5 acres has the potential of producing a crop yield equivalent to 2,400 acres of traditional horizontal farming. In terms of ratio, 1 high-rise farm is equal to 480 traditional horizontal farms [39]
Production of organic crops	Increase organic crops production in a large-scale manner. Adoption of this technology cause reduction of chemical pesticides usage as in conventional practice. The production is ten times more productive as compared to the conventional soil-based farm
Conservation and recycling of natural resources	Implementing hydroponics and aeroponics which consumes a lower amount of water as compared to conventional agriculture. Beneficial in term of water utilization, whereby recycling of water resources can be done. Urban sewage waste can be used as compost and is considered an integrated agricultural practice Ecological services derived include soil forming and conditioning, soil stabilization, waste recycling, carbon sequestration, nutrients cycling, predation, pollination, and habitats [39]
Environment-friendly	Minimize the dependency on land resources and help in regrowth of forests. Due to less use of operational mode and energy, it will lead a decrease in CO_2 emission, thus contributes to environment's wellbeing. This technique offer a long-term effect of ecological balance to prevent soil sterility or pest problems. A lower amount of pesticides help in maintaining and recolonizing beneficial species in including wild flora and fauna [37]
Sustainable urban growth	Applied with a holistic approach in combination with other technologies will able to produce sufficient food supply in order to support the growth of urban population

4.5.3 Limitations in Vertical Farming

Key issues of vertical farming implementation are as follows:

1. Labor intensive.
2. Inputs are unpredictable in term of quantity and time factor.
3. Basic standardized of agronomic practices for vertical farming is not fully implemented and there is a minimal understanding among the farmers.
4. Different weather conditions in different regions of the world causes difficulties in adoption of standardized practices.
5. Lack of varieties for crop production. This aspect needs immediate attention from the researchers, as in the absence of suitable varieties adoption of this technique will be difficult.
6. Lack of knowledge and skills for farming practices in urban populations need to be addressed.

4.5.4 The Implementation of Green Urea in Vertical Farming

Achieving sufficiency in food production along with the conservation of the environment is a major objectives of the agriculture. The constraints in achieving this task are: limiting land and water resources along with the degradation of environmental health due to excessive use of chemicals for nutrition and pest management in agriculture. Anthropogenic activities for development have resulted in a further degradation of the environment and have spurred the rate of climate change. Global efforts have been initiated to reduce the effects of climate change on earth in general and on agriculture in particular. These steps should be complimented with the innovations in production techniques employed in agriculture, and also the scientific utilization of indigenous knowledge which has proven as more sustainable.

The use of green urea technology in the concept of vertical farming can be integrated in the current agriculture system without many disturbances and can be adopted on a global scale. Vertical farming has great potential considering the falling land resources used for agriculture due to extensive urbanization and increase in per capita income in developing nations. The methodology, though a new concern for developing nations, is of great promise and can efficiently meet the challenge in terms of quantity, quality, and variety. Green urea implementation in agriculture requires further exploration and integration of scientific knowledge for incorporation in the main stream agriculture to feed the growing demands. This is a promising possibility for global agriculture and requires further exploration in terms research and integration in main stream for our future sustainability. Rapid change in demographic trends and technological advancements are delivering new innovations in the field of agriculture. These emerging technologies are required to be used wisely to meet the growing demands from modern agriculture. The concept of vertical farming and green urea implementation can be adopted as the viable alternatives for the conventional agriculture to meet the changing demands and needs of mankind. Further, constraints in adoption of such practices should be addressed and linkages between researchers and farmers should be created for suitable measures.

4.6 Summary

Nutrient use efficiency largely depends on the types of fertilizer used and the mode of management practice of fertilizers. Green Urea can be regarded as a complete solution for NUE for crop production system. Exploration of its efficiency through an ammonia production process which utilizes less energy and coating technology to increase the yield and minimize environmental problems are the key factor of its NUS. Its efficiency in spinach in terms of increase of vegetative growth could provide a fundamental understanding about the utilization in other crop systems and in field application. In order to achieve this, optimization techniques on the rate, mode of delivery, and interval time of application are necessary for all types of crop. In addition, integration with various coating materials and technologies is also essential to ensure better NUE. The implementation of Green Urea with various farming systems such as vertical farming associated with internet of things would able to revolutionize agriculture industry.

References

1. Erisman, J. W., Sutton, M. A, Galloway, J., Klimont, Z., & Winiwarter, W. (2008). How a century of ammonia synthesis changed the world. *Nature Geoscience*, 636–639.
2. Sempheho, S. I., Kim, H. T., Mubofu, E., & Hilonga A. (2014). Metic ulous overview on the controlled release fertilizers. *Advances in Chemistry*.
3. Ukessays, (2013). Controlled release fertilizers and nanotechnology traces biology essay http://www.ukessays.com/essays/biology/controlled-release-fertilizers-and-nanotechnology-traces-biology-essay.php.
4. Alexandratos, N., Bruinsma, J. (2012). World agriculture towards 2030/2050: the 2012 revision. ESA working paper No. 12–03. Rome: FAO.
5. IFA (2016). In Harold F. Reetz, Jr (Ed.) *Fertilizers and Their Efficient Use*. Paris, France: Inter National Fertilizer Industry.
6. Baligar, N., & Fageria (2008). *Nutrient Use Efficiency in Plants: An Overview* (pp 1–14) (*Nutrient Use Efficiency from Basic to Advances*).
7. Wiesler, F. (1997). Agronomical and physiological aspects of ammo nium and nitrate nutrition of plants. *Journal of Plant Nutrition and Soil Science*, pp 227–238.
8. Naz, M. Y. & Sulaiman, S. A. (2016). Slow release coating remedy for nitrogen loss from conventional urea: A Review. *Journal of Controlled Release*, pp 109–120.

9. IFA. (2009). *The Global '4R' Nutrient Stewardship Framework: Developing Fertilizer Best Management Practices for Delivering Economic, Social and Environmental Benefits*. Paris, France: International Fertilizer Industry Asso ciation.
10. Dapeng, W. (2016). Technical Summary of Global Enhanced Efficient Nitrogen Fertilizers. Agronews: http://news.agropages.com.
11. AAPFCO. (1997). Association of American Plant Food Control Officials (AAPFCO), Official Publication No. 50, T-29, AAPFCO Inc., West Lafa yette, Indiana,http://www.aap fco.org.
12. Trenkel, M. E. (2010). *Slow and Controlled Release and Stabilized Ferti lizers. An Option for Enchancing Nutrient Use Efficiency in Agriculture*. International Fertilizer Industry Association (IFA), Paris: France.
13. Shaviv, A. (1996). Plant response and environmental aspects as affected by rate and pattern of nitrogen release from controlled release N fertilizers. In *Progress in Nitrogen Cycling Studies* (pp 285–291). Springer.
14. Azeem, B., Ku Shaari, K. Z., Man, Z. B., & Trinh T. (2014). Review on materials and methods to produce controlled release urea fertilizer. *Journal of Controlled Release*, 11–21.
15. Lammel, J. (2005). Cost of the different options available to the farmers: Current situation and prospects. In *IFA International Workshop on En hanced-Efficiency Fertilizers, Frankfurt*. Paris, France: International Fertilizer Industry Association.
16. Shoji, S. (2005). Innovative Use of Controlled Availability Fertilizers with High Performance for Intensive Agriculture and Environmental Conservation. *Science in China Ser. C: Life Sciences*, 912–920.
17. Shaviv, A. (2005). Controlled release fertilizers. In *International Fertilizer Industry Association, International Workshop on Enhanced–Efficiency Fertilizers*. Frankfurt, Paris, France.
18. Trenkel, M. E. (1997). Improving fertilizer use efficiency. In *Controlled Release and Stabilized Fertilizers in Agriculture*. Paris: The International Fertilizer Industry Association (IFA).
19. Lubkowski, K. (2014). Coating fertilizer granules with biodegradable ma terials for controlled fertilizer release. *Environmental Engineering and Management Journal*, 2573–2581.
20. Blaylock, A. (2010). *Enhanced Efficiency Fertilizers*. Colorado State University Soil Fertility Lecture, Agrium Advance Technologies, Love land 10.
21. Shaviv, A. (2001). Proceedings 469. *Improvement of fertilizer efficiency—Product Processing, Positioning and Application Methods* (pp. 23). New York, UK: International Fertiliser Society.
22. Goertz, H. M. (1995). Technology developments in coated fertilizers. In *Proceedings: Dahlia Greidinger Memorial International Workshop on Controlled/Slow Release Fertilizers*. Haifa, Israel: Technion—Israel Institute of Technology.
23. Castro-Enríquez, D. D., Rodríguez-Félix, F., Ramírez-Wong, B., Torres-Chávez, P. I., Castillo-Ortega, M. M., Rodríguez-Félix, D. E., et al. (2012). Preparation, characterization and release of urea from wheat gluten electrospun membranes. *Materials*. (5), 2903–2916.
24. Shi, Y., Zhan, X., Luo, Z., Zhang, Q., & Chen, F. (2008). Quantitative IR characterization of urea groups in waterborne polyurethanes. *Journal of Polymer Science: Part A Polymer Chemistry, 46,* 2433–2444.
25. Rinaudo, M. (2006). Chitin and chitosan: properties and applications. *Progress in Polymer Science, 31,* 603–602.
26. Weis, D. D., & Ewing, G. E. (1996). Infrared spectroscopic signatures of $(NH_4)_2SO4$ aerosols. *Journal of Geophysical Research, 18,* 18709–18720.
27. Martin, P. J., & Jaklevic, J. M. (1986). *Fourier Transform Infrared Spectroscopy (FTIR) Ammonium Sulfate Analysis on Teflon Air Filters*. NC: Raleigh.
28. Sharma, J., & Agarwal, S. (2014). Impact of organic fertilizers on growth, yield, and quality of spinach. *Indian Journal of Plant Science, 3,* 2319–3824.
29. Toledo, M. E. A., Ueda, Y., Imahori, Y., & Ayaki, M. (2003). L-ascorbic acid metabolism in spinach (Spinacia oleracea L.) during postharvest storage in light and dark. *Postharvest Biology and Technology, 28,* 47–57.

30. Salleh, M. M. (2017). Tropical fruit and vegetable production in Malaysia. Malaysian Agriculture Research and Development Institute (MARDI). www.fao.org/fileadmin/templates/.../Tropical_fruit_vegetable_prod_Malaysia.
31. Borris, H., & Kreith, M. (2006). *Commodity Profile: Spinach*. University of California: Agriculture Marketing Resource Centre.
32. Alexandratos N, Bruinsma J. (2012). World agriculture towards 2030/2050: the 2012 revision. ESA working paper No. 12–03. Rome, FAO.
33. Knudsen, M. T., Halberg, N., Olesen, J. E., Byrne, J., Iyer, V., et al. (2005). Global trends in agriculture and food systems. In N. Halberg et al. (Eds.), *Global Development of Organic Agriculture: Challenges and Promises*. UK: CAB International.
34. Gosling, P., Hodge, A., Goodlass, G., & Bending, G. D. (2006). Arbuscular mycorrhizal fungi and organic farming. *Agriculture, Ecosystems and Environment, 113*(1–4), 17–35.
35. Stockdale, E. A., Lampkin, N. H., Hovi, M., Keatinge, R., Lennartsson, E. K. M. (2002). Organic Farming, Old Pond, Ipswich, England. 10. Agronomic and environmental implications of organic farming systems. Advances Agronomy 70: 261–327. (2001).
36. Biao X., Wang X., Ding Z., Yang Y. (2003) Critical impact assessment of organic agriculture. *Journal of Agricultural and Environmental Ethics,* 16(3), 297–311. (12. A very A (2007) Going organic Crops and Soils. *American Society Agronomy*, 40(1), 8–12).
37. Gosling, P., & Shepherd, M. (2005). Long-term changes in soil fertility in organic arable farming systems in England, with particular reference to phosphorus and potassium. *Agriculture, Ecosystems and Environment, 105*(1–2), 425–432.
38. Willer H., Lernoud J. (2014). The world of organic agriculture: Statistics and emerging trends in 2014. (15th ed., pp. 25–32). Switzerland.
39. FAO. (1998). Evaluating the potential contribution of organic agriculture to sustainability goals. Environment and Natural Resources Service. Sustainable Development Department. In *FAO's technical contribution to IFOAM's Scientific Conference, Argentina.*
40. Hergert, G. R., Ferguson, C., Wortmann, C., Shapiro, C., & Shaver, T. (2011). Enhanced efficiency fertilizers: will they enhance my fertilizer efficiency. Proceedings of the 3rd Annual Crop Production Clinics, University of Nebraska-Lincoln Extension.

Chapter 5
Agricultural 4.0: Its Implementation Toward Future Sustainability

5.1 Introduction

Emerging technologies, which sometimes be disruptive, are growing at an exponential rate advancement and progress in technology deemed crucial particularly if it is related to game-changing technology. The Fourth Industrial Revolution has been a subject of interest among many people, particularly in industries, higher learning institutions, governments and research institutions across the globe. It is basically a concept and development that has fundamentally reshaped the way we live. The development is so fast and important that it could impact the society and economy. Recently, we know how some advancement in technologies had changed our diets, lifestyle, eating habit, and our perception and understanding on some issues in regard to the environment. Revolution is defined as a forcible overthrow of a government. In the context of the Industrial Revolution, it is perceived as the forcible overthrow of a current technology. However, it is not only considered disruptive, but it is also very much strongly integrated with other technologies as well.

5.2 History of Industrial Revolution

The Industrial Revolution, which occurred during the eighteenth and nineteenth centuries, was a period during which predominantly agrarian, rural societies in Europe and America became industrial and urban. Prior to the Industrial Revolution, which began in Britain in the late 1700s, manufacturing was often done in people's homes, using hand tools or basic machines [1]. Industrialization marked a shift to powered, special-purpose machinery, factories, and mass production. The iron and textile industries, along with the development of the steam engine, played central roles in the Industrial Revolution, which also saw improved systems of transportation, communication, and banking. While industrialization brought about

N. Yahya, *Green Urea*, Green Energy and Technology,
https://doi.org/10.1007/978-981-10-7578-0_5

an increased volume and variety of manufactured goods and an improved standard of living for some, it has also often resulted in grim employment and living conditions for the poor and working classes.

Before the advent of the Industrial Revolution, most people resided in small, rural communities where their daily existences revolved around farming. Life for the average person was difficult, as income was meager, and malnourishment and disease were common. People produced the bulk of their own food, clothing, furniture, and tools. Most manufacturing was done in homes or small, rural shops, using hand tools or simple machines.

A number of factors contributed to Britain's role as the birthplace of the Industrial Revolution [1]. For one, it had great deposits of coal and iron ore, which proved essential for industrialization. Additionally, Britain was a politically stable society, and was the world's leading colonial power at the time, which meant its colonies could serve as a source of raw materials, as well as he a marketplace for manufactured goods.

As the demand for British goods increased, merchants needed more cost-effective methods of production, which led to the rise of mechanization and the factory system. The textile industry, in particular, was transformed by industrialization. Before mechanization and factories, textiles were made mainly in people's homes (giving rise to the term cottage industry), with merchants often providing the raw materials and basic equipment, and then picking up the finished product. Workers set their own schedules under this system, which proved difficult for merchants to regulate and resulted in numerous inefficiencies. In the 1700s, a series of innovations led to ever-increasing productivity, while requiring less human energy. For example, in 1764, Englishman James Hargreaves (1722–1778) invented the spinning jenny, a machine that enabled an individual to produce multiple spools of threads simultaneously [2]. By the time of Hargreaves' death, there were over 20,000 spinning jennys in use across Britain. The spinning jenny was improved upon by British inventor Samuel Compton's (1753–1827) spinning mule, as well as later machines [2]. Another key innovation in textiles, the power loom, which mechanized the process of weaving cloth, was developed in the 1780s by English inventor Edmund Cartwright (1743–1823) [2].

Development in the iron industry also played a central role in the Industrial Revolution. In the early eighteenth century, Englishman Abraham Darby (1678–1717) discovered a cheaper, easier method to produce cast iron, using a coke-fueled (as opposed to charcoal-fired) furnace. In the 1850s, British engineer Henry Bessemer (1813–1898) developed the first inexpensive process for mass-producing steel [1]. Both iron and steel became essential materials, and were used to produce everything from appliances, tools, and machines to ships, buildings, and infrastructure.

The steam engine was also integral to industrialization. In 1712, Englishman Thomas Newcomen (1664–1729) developed the first practical steam engine (which was used primarily to pump water out of mines) [2]. By the 1770s, Scottish inventor James Watt (1736–1819) had improved on Newcomen's work, and the steam engine went on to power machinery, locomotives, and ships during the Industrial Revolution [3].

The transportation industry also underwent significant transformation during the Industrial Revolution. Before the advent of the steam engine, raw materials and finished goods were hauled and distributed via horse-drawn wagons, and by boats along canals and rivers. In the early 1800s, American Robert Fulton (1765–1815) built the first commercially successful steamboat, and by the mid-nineteenth century, steamships were carrying freight across the Atlantic [3]. As steam-powered ships were making their debut, the steam locomotive was also coming into use. In the early 1800s, British engineer Richard Trevithick (1771–1833) constructed the first railway steam locomotive. In 1830, England's Liverpool and Manchester Railway became the first to offer regular, timetabled passenger services. By 1850, Britain had more than 6,000 miles of railroad track. Additionally, in 1820, Scottish engineer John McAdam (1756–1836) developed a new process for road construction. His technique, known as macadam, resulted in roads that were smoother, more durable, and less muddy [2].

Communication became easier during the Industrial Revolution with inventions such as the telegraph. In 1837, William Cooke (1806–1879) and Charles Wheatstone (1802–1875), patented the first commercial electrical telegraph. By 1840, railways were a Cooke–Wheatstone system, in 1866, a telegraph cable was successfully laid across the Atlantic. The Industrial Revolution also saw the rise of banks and industrial financiers, as well as a factory system dependent on owners and managers. A stock exchange was established in London in the 1770s; the New York Stock Exchange was founded in the early 1790s. In 1776, Scottish social philosopher Adam Smith (1723–1790), who is regarded as the founder of modern economics, published "The Wealth of Nations" [1, 2, 3]. In it, he promoted an economic system based on free enterprise, the private ownership of means of production, and lack of government interference.

The Industrial Revolution brought about a greater volume and variety of factory-produced goods and raised the standard of living for many people, particularly for the middle and upper classes. However, life for the poor and working classes continued to be filled with challenges. Wages for those who labored in factories were low, and working conditions could be dangerous and monotonous. Unskilled workers had little job security and were easily replaceable. Children who were part of the labor force often worked long hours and were used for highly hazardous tasks such as cleaning the machinery. In the early 1860s, an estimated one-fifth of the workers in Britain's textile industry were younger than 15. Industrialization also meant that some craftspeople were replaced by machines [3]. Additionally, urban, industrialized areas were unable to keep pace with the flow of arriving workers from the countryside, resulting in inadequate, overcrowded housing and polluted, unsanitary living conditions in which disease was rampant. Conditions for Britain's working class gradually improved by the later part of the nineteenth century, as the government instituted various labor reforms, after which workers gained the right to form trade unions.

The British enacted legislation to prohibit the export of their technology and skilled workers. However, they had little success in this regard. Industrialization spread from Britain to other European countries, including Belgium, France, and

Germany, and eventually to the USA. By the mid-nineteenth century, industrialization was well-established throughout the western part of Europe and America's northeastern region. By the early twentieth century, the USA became the world's leading industrial nation [3].

The First Industrial Revolution in 1748 occurred when mechanical revolution was powered by water and steam. Steam engines are external combustion engines where the working fluid is separated from the combustion products [1]. Thomas Newcomen invented an atmospheric engine in 1712, the first commercial engine that uses piston. During the first Industrial Revolution, the steam engine replaced water and wind power, and becoming the dominant source of energy in the late nineteenth century. James Watt invented a steam engine in 1781 that brought the First Industrial Revolution. Watt introduced a design enhancement based on Thomas Newcoman's separate condenser, which able to give radical improvement on the power and increased efficiency and improved cost effectiveness of the steam engine, as illustrated in Figs. 5.1 and 5.2.

The second revolution took place with the introduction of the conveyor belt, and hence, mass production was made possible. The conveyor belt is a carrying medium of a belt conveyor system. It consists of two or more pulleys with an endless carrying medium, and a conveyor belt that rotates about the pulleys. The pulleys are powered by moving the belt which moves the materials forward. In 1901, Sandvik invented and started the production of steel-conveying belts. The first conveyor belt for the coal mining industry was invented by Richard Sutcliffe. In 1913, Henry Ford introduced the first moving assembly line for mass production which enables a reduced time from 12 h to 30 min to build a car [2]. This has revolutionized the automotive

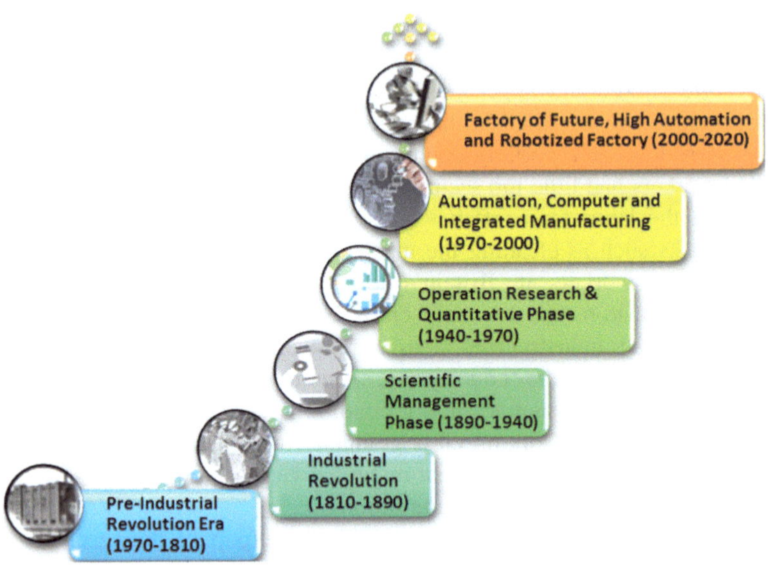

Fig. 5.1 Industrial Revolution history (1970–2020)

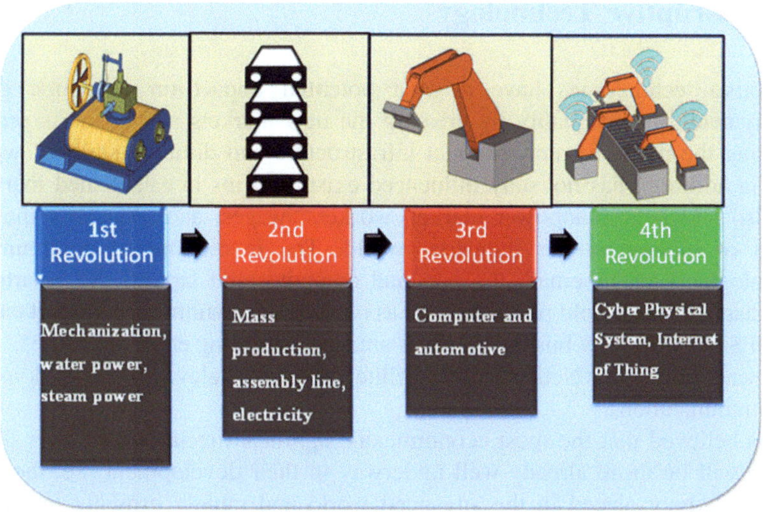

Fig. 5.2 Transformation from industrial 1.0–4.0

industry. The Information Age is a period in human history that is due to the shift of traditional industry that Industrial Revolution brought through industrialization to an economy based on information computerization [3] technology. The digital automation of production by means of electronic and information technology marks the Third Industrial Revolution. On the other hand, the dawn of the Fourth Industrial Revolution began when industry was impacted by the integration of physical and cyber systems, referred to as cyber-physical systems (CPSs). Consequently, the integrated system, production, sustainability, and customer satisfaction formed the intelligent network systems and processes [4]. The Internet of Things (IoT) had greatly changed the landscape of the entire industry. Artificial intelligence which will provide the brain of the entire industry through robotics, self-driven cars, drones, etc., is developed based on industrial needs. New business services are being developed with the intention to improved services for customer satisfaction and to improve productivity and sales by means of connectivity between cyber and physical system. It should be highlighted that the brain of a digital processes is due to advancement of the microelectronics industry. A new phase of robots is currently present in a large number hence, the industry without robotics is almost unconceivable. Robots were designed to execute a dangerous, mass production, and dull tasks, some of each may not be able to be executed by man. They are built to perform some predefined tasks, and their functionality supports business processes [5].

5.3 Disruptive Technology

Disruptive technologies have a great potential impact on economic growth, employment, and inequality by creating the new markets and business practices, revealing the needs for new product infrastructure and different skills of workers. This, in addition, has not only influenced existing firms in established markets; it can also affect the manpower market, workers' wages, and ultimately the distribution of income in the industrial world. The revolutionized communication technologies such as email, the personal computer and laptop, and smartphones have displaced many old products such as typewriters, mainframes, pocket cameras, and GPS devices. New business models are also disrupting entire industries, such as Uber with taxi cabs, Netflix with satellite and cable television, and Skype with telecommunications.

It is believed that the most economically significant technologies over the next decade will be those already well underway in their development, i.e. the mobile Internet, largely placed in the advanced world and rapidly growing in emerging markets; the automation of knowledge work, things such as computerized voices that can handle many customer service calls; the "Internet of Things," such as embedding sensors in physical objects to monitor the flow of products through a factory; and cloud computing. Each of these areas of innovation will contribute an estimated $1 trillion to the world's economy by 2025, even on the low side of their range.

Disruptive technologies can certainly benefit consumers by providing cheaper, more accessible goods or services. Firms that are slow to anticipate or react to disruptive forces might suffer declines in shareholder value and eventually markets. The knock-on effect on manpower markets is more unsettling as workers are often less well-placed to retrain, retool, or relocate, and traditional program of adjustment assistance has proven to be largely ineffective [1].

The 12 areas that exhibit the greatest economic impact and potential to cause disruption by 2025 are: mobile Internet, automation of knowledge work (artificial intelligence), the Internet of Things, cloud technology, advanced robotics, autonomous and near-autonomous vehicles, next generation genomics, energy storage, 3-D printing, advanced materials, advanced oil and gas explorations, and renewable energy [2]. These trends were chosen using four criteria including high rate of technological change, broad potential scope of impact, large economic value affected, and potential for disruptive economic impact. Together, they are estimated to affect trillions of dollars of economic activity and tens of millions of workers.

These disruptive trends are leading in a new era in which digital technologies are meeting or surpassing the capabilities of humans, even in tasks which do not follow a straightforward application of existing rules and were impossible to automate before, such as those involving communication or pattern recognition in uncertain or changing environments. An increase in computer awareness and its capabilities have influenced the need for the processing power of computers to double every two years. This means that there is an exponential growth in computing capability, and that computers of much lower processing power will become cheaper quickly, which then creates the availability of more affordable computers in the market.

By 2025, the mobile Internet, the automation of knowledge work, and the Internet of Things have the potential to deliver an economic value of up to $33 trillion a year worldwide. The estimated range of the impact of dozen technologies is also quite wide, ranging from $14 to $33 trillion by 2025. The technology in renewable energy, advanced oil and gas exploration, and recovery will have a massive effect on the measured impact of those technologies as the energy prices can fluctuate widely.

The optimization of technology will be crucial for total 12 independent, including automation of knowledge work, advanced robotics, next-generation genomics, and Internet of Things, which involves embedding sensors, smart software and communications capability into machines and other physical objects.

The estimation of the transformation of economic growth is important to determine the major forces shaping our technology advancement in the future. Those factors depend on the technological innovations which then turn out to have major economic consequences. The results have some important implications on how we think about innovation. The technologies that will have the greatest impact on the economic potential are the ones that have been evolving for many years in new ways. The technologies that are already in the advanced stage have broad potential impact and could have significant economic impact [3].

For example, in IT technology, computer speed and memory have increased over the past 30 years since the first personal computers were introduced in the market, and fields such as biotechnology are also advancing rapidly. First such as cloud computing and mobile Internet connectivity are not brand-new inventions, as they have existed for years. But the innovations that will flow from those fields, resulting in economic growth, are only the beginning.

Technology advancements have determined a way to put a dollar figure to every technology imaginable all the way out to 2025. Mobile Internet is the single most valuable and disruptive technology we know today by far. With the potential to generate nearly $11 trillion in economic impact, the mobile internet is developing at an alarming rate. (Tables 5.1, 5.2 and 5.3).

In any field, businesses are putting in the same effort trying to mix all these technologies to create profitable revenue. The technologies that have been around for ages and achieved maturity are those that contribute to the real economic benefits of its technology innovation.

Surprisingly, the overwhelming value of mobile Internet is at a very low place given to energy, whether in at a form of fossil fuels or renewables. This is based on

Table 5.1 Top 12 ranking of disruptors technologies [6]

Ranking number	Technology	Potential economic impact ($ trillion)
1	Mobile internet	3.7–10.8
2	Automation of knowledge work	5.2–6.7
3	Internet of Things	2.7–6.2
4	Cloud	1.7–6.2
5	Advanced robotics	1.7–4.5
6	Autonomous and near-autonomous vehicles	0.2–1.9
7	Next-generation genomics	0.7–1.6
8	Energy storage	0.1–0.6
9	3D printing	0.2–0.6
10	Advanced materials	0.2–0.5
11	Advanced oil and gas exploration and recovery	0.1–0.5
12	Renewable energy	0.2–0.3

the significant advancements in an industry that contributes over $1.1 trillion to the economy every year. The impact of technology in this sector is limited by the value of additional output that could cost-effectively be produced. The technology advancement in the field of energy, relative to the economic potential, the renewable energy is the most overhyped technology.

The five technologies that nearly made their list of the most disruptive and impactful technologies are as follows [6]:

1. **Next-generation nuclear**—unlikely to generate a significant impact by 2025;
2. **Fusion power**—unlikely to mature in the report's time frame;
3. **Carbon sequestration**—cost constraints remain high;
4. **Advanced water purification**—more cost-effective approaches are not scalable yet;
5. **Quantum computing**—too much uncertainty about its applicability and impact.

By the time, these disruptive potential technologies exert their influence on the economy in 2025 (Table 5.2); it will be too late for business leaders, policy makers, and stakeholders [7]. They must look ahead, identify the technologies that could affect them, and determine how to shape markets and policies in ways that will serve their main interests. These technologies could fuel a decade of rapid innovation in products, services, business processes, and go-to-market strategies. Companies will have new ways of developing and producing products, organizing their businesses, and reaching consumers and business-to-business customers. Business leaders will need to determine when, how, and whether to take advantage of new technologies and be prepared to move quickly when others use emerging technologies to mount challenges.

Table 5.2 Top six disruptors technologies [6, 8]

Disruptors technology	Effect	Capability
Mobile internet	Increasingly inexpensive and capable mobile computing devices and Internet connectivity	Enabling more efficient delivery of services and opportunities to increase workforce productivity
Automation of knowledge work	Intelligent software systems that can perform knowledge work tasks involving unstructured commands and subtle judgments	A threat to the service sector, especially with voice recognition allowing computers to interact with customers. It will allow for automation of a lot of knowledge work and make it cheaper and more accessible
Internet of Things	Networks of low-cost sensors and actuators for data collection, monitoring, decision making, and process optimization	A positive for industry because it allows companies to manage assets and optimize performance of production process by having improved sensors and remote monitoring
Cloud computing	Use of computer hardware and software resources delivered over a network or the Internet, often as a service	Negates the need for having costly hardware equipment because software and hardware are now accessible remotely over the Internet
Advance robotics	Increasingly capable robots with enhanced senses, and intelligence used to automate tasks or augment humans	It is an obvious threat to manufacturing jobs, but can boost production and reduce costs and also a threat to industries such as health care as some tasks would be able to be performed by robots, and services such as cleaning and maintenance
Autonomous and near-autonomous vehicles	Vehicles that can navigate and operate with reduced or no human intervention. If regulation allows, as early as 2020 autonomous cars, aircraft, and boats can revolutionize transportation	

In the twenty-first century, all business leaders must understand technology. They must develop their own well-informed views of what developments such as cloud computing could do for their enterprises and work to separate hype from reality. Leaders should think carefully about how specific technologies could drive economic impact and disruption in ways that could affect their businesses [8]. Leaders should invest in their own technology knowledge, not to become

Table 5.3 Development characteristics and key features of target country [7, 9, 12]

Country/Target	Development characteristic	Key features
Taiwan **Target**: To promote Taiwan as a hub for Agriculture 4.0	• Agriculture 4.0 • Smart farming devices • Large-scale precision agriculture projects	• Precision farming • Agriculture 4.0 • Drones, robotics • Internet of Things (IoT) • Vertical farms • Artificial intelligence (AI) • Solar energy
Thailand **Target**: Aimed to pull the country out of the middle income trap. (average annual income of farmers from 56,450 baht to 390,000 baht within the next 20 years)	• Thailand 4.0 and Agriculture 4.0 • Reducing inefficiencies, water consumption, fertilizers insecticides, and other chemicals and harmful effects on soil, animals, and people • Give way to foreign investors to make Thailand 4.0 and Agriculture 4.0 more feasible	• Farmers actively improved the qualities of fruit varieties in response to consumer demand • Local firms were able to adapt imported machinery to match local farm conditions and further export the modified versions to other low-income countries • Hired services and rental markets, particularly for combined harvesters, emerged and have since developed into a full-fledged practice that allows farmers to share the costs • Multistakeholder partnership projects • Investment programs that will bring together the non-government organizations, companies, the government, universities, and farmers • Technological transformation that will also provide access to shared information, technology, finances, and markets
Vietnam **Target**: Vietnam has emerged as one of the world's leading exporters of agricultural commodities and is among the top five for aquatic products, rice, coffee, tea, cashews, black pepper, rubber, and cassava	• Agriculture 4.0 • High-tech and clean agricultural projects • Developing high-tech agriculture is an important task and also the inevitable trend of Vietnam's socioeconomic development strategies in the context of deeper international integration	• The development of high-tech agriculture at lower market rates • Preferential loans to high-tech and clean agricultural projects. Interest rates of the loans will be 0.5–1.5% per year lower than other average lending rates

(continued)

Table 5.3 (continued)

Country/Target	Development characteristic	Key features
British **Target**: New patterns of crop rotation and livestock utilization paved the way for better crop yields, a greater diversity of wheat and vegetables, and the ability to support more livestock	• Agriculture 4.0 • The increased availability of farmland • A favorable climate • More livestock • Improved crop yield	• Norfolk four-course crop rotation: fodder crops, particularly turnips and clover, replaced leaving the land fallow • The Dutch improved the Chinese plow so that it could be pulled with fewer oxen or horses • The removal of common rights to establish exclusive ownership of land • Development of a national market free of tariffs, tolls, and customs barriers • Transportation infrastructures, such as improved roads, canals, and later, railways • Land conversion, land drains, and reclamation • Increase in farm size • Selective breeding
India **Target**: Efforts on overcoming language barriers and encouraging collective farming to reduce land fragmentation	• Agriculture 4.0 • Enables farmers to sow right crops at the right location in order to get optimized yield in the limited patch of land • Delivers precise information on the use of agri-inputs leading to a reduction in its wastage and hence resulting in considerable savings and lesser reliance on rural credit • Innovative practices will make the farmers less dependent on rains or will educate them on water-harvesting techniques in case they still rely on rains • Data acquisition for precise data accurate up to the remotest village is available on categories like climate, rainfall, land, soil condition	• At 190 million hectares, India holds second largest agricultural land in the world • Around 800 million of rural population depends on agriculture as their primary source of income and livelihood • Using data science to precisely regulate and optimize the agricultural investment • Management measures in accordance with specific condition of each unit of farmland so as to maximize output and economic benefit while reducing use of resources to protect agricultural ecology

programmers or compulsive Facebook posters, but they should keep abreast of technology trends and pay attention to what their most tech-savvy customers are doing and saying.

No.	Value driver	Description
1	Resource/ processes	• Smart energy consumption • Intelligent lots • Real-time optimization – Productivity increase of 3–5%
2	Asset utilization	• Routing flexibility • Machine flexibility • Remote monitoring and control • Predictive maintenance • Augmented reality for MRO[1.] – 30–50% Reduction of total machine downtime.
3	Labor	• Human–robot collaboration • Remote monitoring and control • Digital performance management • Automation knowledge work – 45–55% increased productivity in technical professions through automation of knowledge work
4	Inventory	• In situ 3D printing • Real-time supply chain optimization. • Batch size 1 – Costs inventory holding decreased by 20–50%
5	Quality	• Statistical process control • Advance process control • Digital quality management • Costs for quality reduced by 10–20%
6	Supply/demand matching	• Data-driven demand prediction • Data-driven design to value – Forecasting accuracy increased to 85%+
7	Time to market	• Customer cocreation/open innovation • Concurrent engineering • Rapid experimentation and simulation – 20–50% reduction in time to market
8	Service/after sales	• Predictive maintenance • Remote maintenance • Virtually guided self-service – 10–40% reduction of maintenance costs

Time is crucial and the world is changing as at the highest speed of the Internet, and technology is continually evolving. Strategies can quickly fall behind, so the rhythm of planning must keep in pace. When technologies have disruptive potential, the stakes are even higher and the range of strategic implications is wider. This is a big red flag against the sole focus by companies on their largest, most established markets and related value propositions. In doing so, companies can miss the ways in which disruptive technologies can jump industry or market boundaries and change the rules of the game.

When necessary, leaders must be prepared to disrupt their own businesses and make the investments necessary for changes (Fig. 5.3): As the past two decades have shown, successful companies repeatedly reinvent themselves to keep up. This will require continuous experimentation and investment. Early investment will probably dilute the profitability of a company's portfolio in the near term, but is ultimately tomorrow's sources of growth that ensures the enterprise's future [7]. Companies that reallocate resources early to capture trends often have higher returns and are more likely to survive in the long term. Failing to reinvent and focusing only on existing markets might open the door for disruptor's potential, particularly at the bottom end of the market.

The democratization of technology is currently advancing, reducing the barriers to entry, and allowing entrepreneurs and other new competitors to disrupt well-established markets and industries. Cloud services make it easier for new companies with little capital to obtain operating infrastructure and access to markets that has taken global companies decades to build. 3D printing goes a step further; it

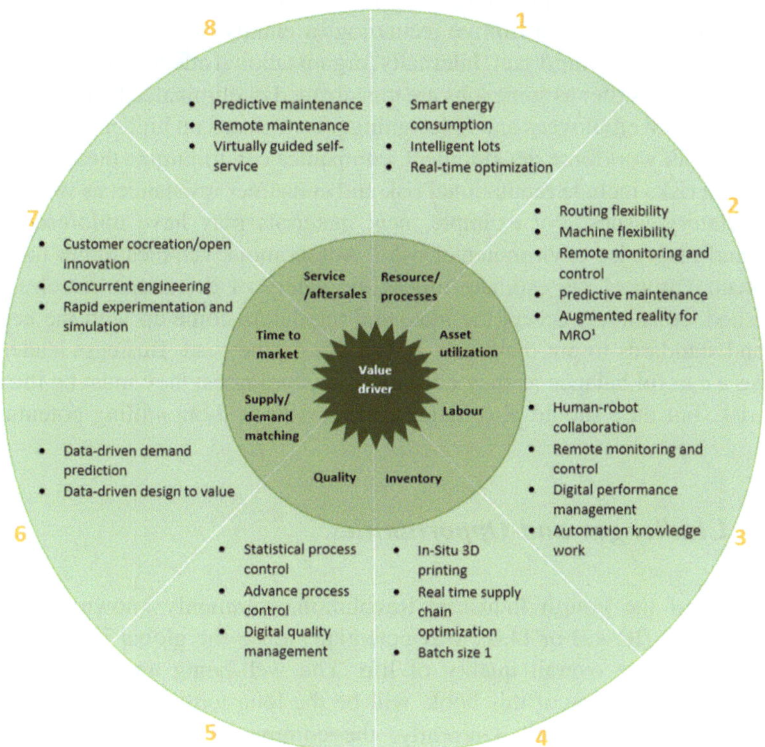

Fig. 5.3 Value driver of disruptive technology [6]

not only opens up markets to competition from entrepreneurs, but it also has the potential to shift value directly to consumers as they learn how to make things that they used to buy [6]. To compete in this environment, companies need deep sources of value or competitive advantage. The prospect of more than two billion new Internet users in developing economies promises both access to new consumers and the threat that those consumers will go into business against you. To survive, companies will need to learn the tricks of the Internet trade and multisided business models such as online advertising or monetizing exhaust data, which need not be reserved for Google and Facebook.

Some of the largest opportunities and challenges for business leaders will arise from new tools that could transform how work is done. With technologies such as advanced robotics and automated knowledge work, companies can potentially have unique opportunities to realize rapid improvements in productivity. These tools could redefine jobs as tasks are augmented by, or transferred to, machines, thus requiring new skills for workforces. Knowledge workers are the foundation of future success, in all sectors of the economy; by 2025, some manufacturers could be hiring more designers and robotics experts than assemblers [7].

Companies that use technology to make knowledge employees more productive will gain a large business model advantages and attract the best talent. Companies will need to have the right people, along with the training systems to keep these workers' skills current. Adopting disruptive technologies entails risks, and managing these risks will be critically important. Internally, organizational effectiveness and cohesion could potentially suffer as some jobs are transformed or eliminated by technology [6]. By working with employees and redesigning jobs to focus on higher value skills and by investing in workforce development, companies can minimize these risks.

External risks include reputational risk and consumer resistance, as well as safety and regulatory issues. For example, new materials may have unforeseen health effects and may pose environmental risks. Autonomous vehicles might not deliver the potential impact we estimate unless the safety of driverless vehicles can be established, consumers accept the idea, and regulators come up with the necessary rules and standards to put these cars and trucks on the road. Business leaders need to strike a careful balance as they adopt new technologies; they must be thoughtful about risk, but they should also manage these risks without stifling potential.

5.3.1 Challenges and Opportunities

The arrival of the Fourth Industrial Revolution, commonly known as Industrial Revolution 4.0 (IR 4.0 or I4.0), will potentially raise the global income and certainly improve the overall quality of life. The well-being and sustainability of human, the main theme of this book, will be the long-term gain, in terms of both efficiency and productivity. Currently, the communication, transportation, and entertainment cost are greatly reduced, and hence, this has affected our lifestyle for the better.

Low growth, volatile energy prices, environmental concerns, and rising expectations from consumers, are some of the challenges the crop production industry is facing today, along with diminishing production profit margins for farmers. To grow food the world needs, crop production systems need innovative solutions to produce more in an environmentally, economically, and socially viable manner.

Research and development in everything from crop development to land usage to pre- and post-harvest crop management will be crucial in providing high-quality, affordable food that protects both natural resources and human health while remaining competitive. Some innovative solutions are already available, and their implementation sometimes made compulsory by law, such as adherence to the Common Agricultural Policy (CAP). But how can we improve access to and use of innovative solutions? What challenges do farmers face as they work to ensure sustainable crop production? Are the existing technologies and tools enough to make crop production sustainable in the long run?

Two main challenges faced by our current food production are to produce food efficiently and safely and contribute to future sustainably. Each day, the world's population grows by 200,000 people, which means many new mouths to feed. The UN estimates that food production will have to be expanded by 70% by 2050. This needed expansion of food production is a daunting challenge, but one that can be met if we make good use of modern technology, and support continued innovation [9].

Agricultural efficiency has already expanded greatly since the 1960s, when the expanded use of new crop protection techniques was accompanied by a huge increase in agricultural capacity. Society is increasingly aware of the dual challenge of feeding a growing population while protecting the environment.

The crop protection practices developed in the 1960s have had a detrimental effect on soil [3]. But different practices can help turn this situation around. The use of cleverly managed, long-term crop rotations, can be used to not only build up soil fertility and health, but also to break cycles of pest outbreaks and disease. One approach to protecting the environment has been organic farming, which is growing in popularity. But organic farming alone may not be able to meet the world's growing food demand. The crop protection industry believes that high-yield pest control can be combined with efficient farming practices to increase future sustainability.

As we know the first industrial revolution relied heavily on human and animals while the arrival of the second revolution had reduces the labor demand and time of production. Since the third and fourth industrial revolutions come with the digitalization era and information age, the blue collar workforce might be reduced greatly, thereby changing the structure of the society. The substitution of automation across the entire economy hence the operating cost and capital investment cost returns might have a new model. Potentially a new rewarding occupation which is health and environment safety and efficient which are to some extent is connected via social media. Some jobs will certainly be vanished.

5.4 Agriculture 4.0

5.4.1 History of Agriculture

The history of agriculture visualized mankind's development and cultivation of producing food, fiber, fuel, and other goods by the systematic raising of plants and animals. Prior to the development of plant cultivation, human beings were hunters and gatherers. The knowledge and skill of learning to care of soil and growth of plants advanced the development of human society, allowing clans and tribes to stay in one location generation after generation. Archeological evidence indicates that such developments occurred 10,000 or more years ago.

Due to agriculture, cities and trade relations between different regions and groups of people developed, further enabling the advancement of human societies and cultures. Agriculture has been an important aspect of economics throughout the centuries prior to and after the Industrial Revolution. The sustainable development of the world's food supplies impacts the long-term survival of the species. There is always a need to ensure that agricultural system is maintain thier green approach with our ecosystem.

By 7000 BCE, sowing and harvesting reached Mesopotamia, in the fertile soil just north of the Persian Gulf, Sumerian ingenuity systematized it and scaled it up. By 6000 BCE [2], farming was entrenched on the banks of the Nile River. During this time, agriculture was developed independently in the Far East, probably in China, with rice as the primary crop [7]. Maize was first domesticated, probably from teosinte, in the Americas around 3000–2700 BCE, though there is some archeological evidence of a much older development. The potato, tomato, pepper, squash, several varieties of bean, and several other plants were also developed in the New World, as was quite extensive terracing of steep hillsides in much of Andean South America. Agriculture was also independently developed on the island of New Guinea.

The reasons for the development of farming may have included climate change, but possibly there were also social reasons such as accumulation of food surplus for competitive gift-giving as in the Pacific Northwest potlatch culture. There was a gradual transition from hunter-gatherer to agricultural economies after a lengthy period during which some crops were deliberately planted and other foods were gathered in the wild. Although localized climate change is the favored explanation for the origins of agriculture in the Levant, the fact that farming was "invented" at least three times elsewhere, and possibly more, suggests that social reasons may have been instrumental.

Full dependency on domestic crops and animals did not occur until the Bronze Age, by which time wild resources contributed a nutritionally insignificant component to the usual diet. If the operative definition of agriculture includes large-scale intensive cultivation of land, monocropping, organized irrigation, and use of a specialized labor forces, the title "inventors of agriculture" would fall to the Sumerians, starting ca. 5,500 BCE. Intensive farming allows for a much greater density of population that can be supported by hunting and gathering, and enables the accumulation of excess products for off-season use, or to sell/barter. The ability of farmers to feed large numbers of people whose activities have nothing to do with material production was the crucial factor in the rise of standing armies. Sumerian agriculture supported a substantial territorial expansion, together with much internecine conflict between cities, making them the first empire builders. Not long after, the Egyptians, powered by farming in the fertile Nile valley, achieved a population density from which enough warriors could be drawn for a territorial expansion more than tripling the Sumerian empire in area. [5]

The invention of a three-field system of crop rotation during the Middle Ages, and the importation of the Chinese-invented moldboard plow, vastly improved agricultural efficiency. After 1492, the world's agricultural patterns were shuffled in the widespread exchange of plants and animals known as the Columbian Exchange [10]. Crops and animals that were previously only known in the Old World were now transplanted to the New World and vice versa. Perhaps most notably, the tomato became a favorite in European cuisine, and maize and potatoes were widely adopted. Other transplanted crops include pineapple, cocoa, and tobacco. In the other direction, several wheat strains quickly took to western hemisphere soils and became a dietary staple even for native North, Central, and South Americans [11].

Agriculture was a key element in the Atlantic slave trade, Triangular trade, and the expansion by European powers into the Americas. In the expanding plantation economy, large plantations produced crops that included sugar, cotton, and indigo, which were heavily dependent upon slave labor.

By the early 1800s, agricultural practices, particularly careful selection of hardy strains and cultivators, had improved so much that yield per land unit was many times that seen in the Middle Ages and before, especially in the largely virgin soils of North and South America. The eighteenth and nineteenth centuries also witnessed the development of glass houses or greenhouses, initially for the protection and cultivation of exotic plants imported to Europe and North America from the tropics. Experiments on plant hybridization in the late 1800s yielded advances in the understanding of plant genetics, and subsequently, the development of hybrid crops. Storage silos and grain elevators appeared in the nineteenth century. However, increasing dependence upon monoculture crops leads to famines and food shortages, the most notable of which is the Irish Potato Famine (1845–1849) [9].

The birth of industrial agriculture coincides with that of the Industrial Revolution. With the rapid rise of mechanization in the late nineteenth and twentieth centuries, particularly in the form of tractor, farming tasks could be performed with a speed and on a scale that was previously impossible. These advances, joined with science-driven innovations in methods and resources, have led to efficiencies enabling certain modern farms in the USA, Argentina, Israel, Germany, and a few other nations to output volumes of high-quality product per land unit at what may be the practical limit. The development of rail and highway networks and the increasing use of container shipping and refrigeration in developed nations have also been essential to the growth of mechanized agriculture, allowing for the economical long-distance shipping of product.

The identification of nitrogen and phosphorus as critical factors in plant growth led to the manufacture of synthetic fertilizers, making it possible for more intensive types of agriculture. The discovery of vitamins and their role in animal nutrition in the first two decades of the twentieth century led to vitamin supplements, which in the 1920s allowed certain livestock to be raised indoors, reducing their exposure to adverse natural elements. The discovery of antibiotics and vaccines facilitated raising livestock in larger numbers by reducing disease. Chemicals developed for use in World War II gave rise to synthetic pesticides. Other applications of scientific research since 1950 in agriculture include gene manipulation and hydroponics [9].

Agricultural production across the world doubled four times between 1820 and 1975. Between 1820 and 1920; between 1920 and 1950; between 1950 and 1965; and again between 1965 and 1975, so as to feed a global population of one billion human beings in 1800 and 6.5 billion in 2002 [6]. During the same period, the number of people involved in farming dropped as the process became more automated. In the 1930s, 24% of the American population worked in agriculture compared to 1.5% in 2002; in 1940, each farm worker supplied 11 consumers, whereas in 2002, each worker supplied 90 consumers [8]. The number of farms has also decreased, and their ownership is more concentrated. In the USA, four companies kill 81% of cows, 73% of sheep, 57% of pigs, and produce 50% of chickens, cited as an example of "vertical integration" by the president of the US National Farmers Union [7].

While industrial agriculture strives to lower costs and increases productivity, the methods of industrial agriculture also have unintended consequences. The degree and significance of these unintended consequences are subject to debate, as is the question of the best way to deal with these consequences [12].

The Green Revolution, the worldwide transformation of agriculture that led to significant, increases in agricultural production between the 1940s and 1960s, occurred as the result of programs of agricultural research, extension, and infrastructural development, instigated and largely funded by the Rockefeller Foundation, along with the Ford Foundation, and other major agencies [9]. The Green Revolution in agriculture helped food production to keep pace with worldwide population growth. The projects within the Green Revolution spread technologies that had already existed, but had not been widely used outside of industrialized nations. These technologies included pesticides, irrigation projects, and synthetic nitrogen fertilizer.

The novel technological development of the Green Revolution was the production of what some referred to as a miracle seeds. Scientists created strains of maize, wheat, and rice that are generally referred to as the high yielding varieties (HYVs). HYVs have an increased nitrogen-absorbing potential compared to other varieties. Since cereals that absorbed extra nitrogen would typically lodge, or fall over before harvest, semidwarfing genes were bred into their genomes. Norin 10 wheat, a variety developed by Orville Vogel from Japanese dwarf wheat varieties, was instrumental in developing Green Revolution wheat cultivators. IR8, the first widely implemented HYV rice to be developed by IRRI, was created through a cross between an Indonesian variety named "Peta" and a Chinese variety named "Dee Geo Woo Gen" [13].

HYVs significantly outperform traditional varieties in the presence of adequate irrigation, pesticides, and fertilizers. In the absence of these inputs, traditional varieties may outperform HYVs. One criticism of HYVs is that they were developed as F1 hybrids, meaning they need to be purchased by a farmer every season rather than saved from previous seasons, thus increasing a farmer's cost of production.

The idea and practice of sustainable agriculture have arisen in response to the problems of industrial agriculture. Sustainable agriculture integrates three main goals: environmental stewardship, farm profitability, and prosperous farming communities. These goals have been defined by a variety of disciplines and may be looked at from the vantage point of the farmer or the consumer.

5.4.2 Fourth Industrial Revolution in Agriculture (Agriculture 4.0)

Industrial agriculture is a modern form of farming that refers to the industrialized production of livestock, poultry, fish, and crops. The methods of industrial agriculture are techno-scientific, economic, and political. They include innovation in agricultural machinery and farming methods, genetic technology, techniques for achieving economies of scale in production, the creation of new markets for consumption, the application of patent protection to genetic information, and global trade. These methods are widespread in developed nations and are increasingly prevalent worldwide.

Agriculture is a conservative industry yet it has been a strong pillar of the economy. However, the agriculture sector is facing some challenges, including food security, climate change such as El Nino, water contamination and scarcity, and high energy demand. Similar to other industries, agriculture is affected by the arrival of the Fourth Industrial Revolution as well. Hence, the implementation of advanced technologies such as sensors, smart farming, vertical farming, smart fertigation systems, green methods of application, and green urea are essential to meet the demand of ever-increasing population. Sustainable development with fundamental yet high-level integration of scientific achievement and modern technologies in sectors of digitalization, biotechnology, and nanophysics hence means that the arrival of IR 4.0 (Fig. 5.4) is expected to have a compelling change in the agriculture industry. When Germany first launches the IR 4.0, it did not anticipate that this would change the entire industry. Some advancement in agriculture based on some country, such as Africa, Vietnam, Nigeria, China, and India, will be reported in the following section.

Fig. 5.4 Impact of Agriculture 4.0

Precision engineering is the highlight of the agricultural industry in the next few years. Taiwan is among the few countries, besides Germany that adopts the agriculture in IR 4.0 [14]. Just like any other business, yield of crops will be the ultimate aim and this can be achieved with greater environment prediction. Predictive monitoring systems, especially those related to soil and weather conditions would be essential for the industry. It is full of potential for future farmers that adopt the right components of Agriculture 4.0, such as IoT, smart sensors, robotics, drones and satellites, hydroponics and aquaponics and solar cells among others. Data analytics is important for Agriculture 4.0, as the data carries important information for farmers. Some startups are offering farmers modern equipment in digital form, software, and innovative tools that could capture, monitor, process, and perform real-time monitoring activities [15]. However, it should be noted that since agriculture is a conservative industry, some farmers are receptive in advanced technology which would actually increase their profit to several folds. A drastic shift in strategies by policy makers and the government, is required to ensure that the middle income model of a country such as Vietnam, Malaysia and Thailand is shifted to a high income nation.

5.5 Summary

Advancement in technology is deemed important particularly if it is related to a game-changing technology. Industrial Revolution 4.0 (IR 4.0 or I4.0) will potentially raise the global income and certainly improve overall quality of life. It has been a subject of interest by many people, particularly in industry, higher learning institution, government and research institutions across the world. Developing high-tech agriculture is an important task in the implementation of Agriculture 4.0 and also inevitable trend in the socioeconomic development strategies in the context of deeper international integration. Data science can be used to systematically regulate and optimize agricultural investments. The estimation of the transformation of economic growth is important to determine the major forces that shape our technology advancement in future. These factors depend on technological innovations which then turn out to have major economic consequences. Agriculture 4.0 creates the new patterns of crop rotation, and livestock utilization that pave the way for better crop yields, a greater diversity of wheat and vegetables, and the ability to support more livestock.

References

1. Lewis and Reinhold, (1990). *Roman civilization volume* I. Columbia University Press. ISBN 0231071310.
2. Howatson, M. C. (1989). *The oxford companion to classical literature*. Oxford University Press.
3. Milanovic, B. (2016). *Global inequality: A new approach for the age of globalization*. Harvard University Press.
4. Alexandratos, N., & Bruinsma, J. (2012). World agriculture towards 2030/2050: the 2012 revision. ESA Working paper No. 12–03. Rome, FAO.
5. Kjeilen, T. (2007). Ancient Egypt: Economy. *Lexiorient*. Accesed 13 Aug 2007.
6. Manyika, J., Chui, M., Bughin, J., Dobbs, R., Bisson, P., & Marrs, A. (2013). *Disruptive technologies: Advances that will transform life, business, and the global economy* (Vol. 12). New York: McKinsey Global Institute.
7. Needham, J. (1984). Science and civilization in China: Volume 6, Biology and Biological Technology. Caves Books Ltd.
8. Kanbur, R., & Stiglitz, J. (2015). Wealth and income distribution: New theories needed for a new era. VoxEU, 18 August.
9. Market Wire.(2003). Global Agricultural Issues Take Center Stage at World Agricultural Forum's 2003 World Congress. FindArticles.com. Accessed 13 Aug 2007.
10. Rymer, E. (2007). Farming in Mesopotamia. Accessed 13 Aug 2007.
11. Empires Past: Aztecs: Farming and Agriculture. *ThinkQuest*. (1998). Accessed 13 Aug 2007.
12. Agriculture in the Indus Valley Civilization. *This is My India* (2006). Accessed 13 Aug 2007.
13. Ronald, L. (1997) Plain, Trends in U.S. Swine Industry, (September 24, 1997).
14. Cheng, J. D., Lin, J. P., Lu, S. Y., Huang, L. S., & Wu, H. L. Hydrological characteristics of betel nut plantations on slopelands in central Taiwan/Caractéristiques hydrologiques de plantations de noix de bétel sur des versants du centre Taïwan *53*(6), 1208–1220.
15. Stephen, J. C. (2003). Columbian exchange. Accessed 13 Aug 2007.

MIX
Papier aus verantwortungsvollen Quellen
Paper from responsible sources
FSC® C105338

If you have any concerns about our products,
you can contact us on
ProductSafety@springernature.com

In case Publisher is established outside the EU,
the EU authorized representative is:
Springer Nature Customer Service Center GmbH
Europaplatz 3, 69115 Heidelberg, Germany

Printed by Libri Plureos GmbH
in Hamburg, Germany